Springer
Tokyo
Berlin
Heidelberg
New York
Hong Kong
London
Milan
Paris

M. Kumagai · W.F. Vincent (Eds.)

Freshwater Management

Global Versus Local Perspectives

With 53 Figures

Springer

Michio Kumagai, D.Sc.
Chief Research Scientist
Lake Biwa Research Institute
1-10 Uchidehama
Otsu 520-0806, Japan

Warwick F. Vincent, Ph.D.
Professor
Department of Biology
Laval University
Québec, QC G1K 7P4, Canada

Cover photo: Lake Namucuo is one of the highest lakes in the world with an altitude of 4718 m, a surface area of 1961 km^2, a volume of 76.8 km^3, and a mean depth of 39.2 m. It is located on the Tibetan Plateau and is suffering from salinization due to climate change (*by D. Khadbaatar*).

ISBN 978-4-431-00488-2 Springer-Verlag Tokyo Berlin Heidelberg New York

Library of Congress Cataloging-in-Publication data applied for.

Printed on acid-free paper

Typesetting: Authors

SPIN: 10905678

This volume is dedicated to
the memory of

Professor Ryohei Tsuda

in appreciation of his fundamental research
on Lake Biwa, Japan,
and
his participation
as a leading scientist in BITEX'93.

Preface

The globalization of trade, monetary and fiscal policies, capital markets, and investment patterns is reshaping the world economy and is leading to new financial, commercial, and marketing structures as well as unprecedented economies of scale. Simultaneously, national and international awareness and responses to accelerating environmental degradation continue to strengthen. There is consensus among most developed countries that the rapidly evolving new economic order needs to be well integrated with policies to maintain or restore environmental quality. Many challenges remain, however, in evaluating the geo-ecological implications of economic globalization, and in formulating the appropriate management responses. In lakes and rivers, the management of water supply and quality has largely proceeded on the basis of local considerations rather than at the global scale that has been more typical of environmental management of the atmosphere and ocean. It is increasingly apparent, however, that high-quality water resources are now in critically short supply not only because of local problems such as over-irrigation and eutrophication, but also as a result of larger-scale climate effects on the hydrosphere. This magnitude of impact will increasingly require the integrated monitoring and management of water resources on a planetary scale, with world criteria for environmental assessment, restoration, and conservation strategies. The increasing extent of world trade in potable freshwater heightens the urgency for establishing international approaches, criteria, and regulations. The recent World Conference on Sustainable Development held in Johannesburg, South Africa, in 2002 underscored the critical need for improving both the quantity and quality of global freshwater supplies in many countries of the world.

Although global-scale programmes provide an attractive, long-term solution to environmental management, they may fail to meet local needs and constraints, especially in the short term. The supply and quality of water in lakes and rivers are highly dependent on local biogeochemical, climatic, land-use, and economic factors, and methods of successful water management in one area may be inappropriate or may require substantial modification before they can be applied to another area. It is therefore essential to evaluate the merits and demerits of locally based approaches and policies versus globally integrated and more general strategies for water resource management. In the final analysis, careful decisions based on ecological understanding will be essential for long-term success.

This book was conceived as a forum to explore the dichotomy of global versus local strategies for lake, river, and reservoir management. The project began as 'The Lake Debate', a symposium that took place at the Lake Biwa Research Institute in conjunction with the World Lakes Conference in Otsu, Japan, in October 2001, and that led to the further planning and development of this volume. We invited scientists from throughout the world who are playing leading roles in aquatic research and in water assessment and management, and each individual or research group was asked to emphasize either local or global approaches towards specific types of water issues. The book thus presents a series of contrasting perspectives on each specialist topic. In the final chapter we attempt to integrate these perspectives and to identify strengths, weaknesses, and complementarities of the two approaches that will help refine future strategies for the sustainable use of the world's precious freshwater resources.

We thank the director and staff of the Lake Biwa Research Institute for their valuable support to this project; the Shiga Prefectural Government, the Kansai Research Organization for Hydrosphere Environments, and the Natural Sciences and Engineering Research Council of Canada for funding assistance; and especially the authors and reviewers of each chapter for their contributions to this volume. We are also grateful to Dr. Yasuaki Aota, Ms. Tomoko Naya, and Ms. Masumi Harabe who helped proof checking, printing, and figure reproduction with careful faithfulness.

Michio Kumagai
Warwick F. Vincent

Contents

List of First Authors

Dr. Michio Kumagai (kumagai@lbri.go.jp) is Chief Research Scientist at the Lake Biwa Research Institute, Japan. He organized the international field experiment BITEX '93, which was attended by 177 participants and ran for almost one month at Lake Biwa in 1993. He was also the coordinator for the cyanobacterial risk assessment programme "CRAB" which commenced in 1995 in cooperation with international scientists. He was the first editor-in-chief of *Limnology*, and has been a member of the Limnological Committee, Science Council of Japan since 1997. His present concern is the implementation of a management plan on the future of Lake Biwa.

Dr. Richard Robarts (Richard.Robarts@EC.GC.CA) is Director of the UNEP GEMS/Water Programme Office, the United Nations Environment Programme's global environmental monitoring system for water quality. He is an Adjunct Professor in the Department of Applied Microbiology, University of Saskatchewan; the Department of Biology, University of Alberta and in the Department of Biological Sciences, Napier University, UK. Richard is an Associate Editor of the *Canadian Journal of Fisheries and Aquatic Sciences*, *Aquatic Ecology*, North American Editor for *Ecohydrology & Hydrobiology* and Editor of *SILnews*. He is also a member of the Scientific Board for the International Centre of Ecology, Polish Academy of Sciences in Warsaw, and the Steering Committee of the UNESCO IHP-VI Ecohydrology Project. Dr. Robarts is a member of the Interdisciplinary Committee of the World Cultural Council (Mexico), which nominates individuals for the Albert Einstein World Award (Science) in recognition for their valuable achievements that benefit mankind. He is also a scientific member of the Research Planning Committee, Sustainable Forest Management Networks of Centres of Excellence, University of Alberta and a member of the Advisory Board for the Fourth World Water Forum to be held in Montreal in 2006. Dr. Robarts is President of the Safe Drinking Water Foundation, a registered charitable foundation whose mission is to support research to produce safe drinking water, and to raise public awareness of the importance of safe drinking water in rural areas.

Dr. Jean-Jacques Frenette (Jean-Jacques_Frenette@UQTR.CA) is a Professor in the Department of Chimie-biologie at Université du Québec à Trois-Rivières (UQTR, Trois-Rivières, Canada) and an Adjunct-professor at Université Laval (Québec City, Canada). He is a full member of the Groupe de Recherche Interuniversitaire en Limnologie du Québec (GRIL) and associate member of the Québec Océan oceanographic group. His research interest focuses on the impact of physical factors on primary production and lower food web processes in lake and river ecosystems. Before joining UQTR in 1999 he led research projects at Lake Biwa, Japan, on the mechanisms of bloom-formation by noxious cyanobacteria and the impact of UV radiation on planktonic communities. He now leads a research group of Ph.D., M.Sc., and undergraduate students working on the structure and functioning of the St. Lawrence River ecosystem, Canada. He collaborates with various provincial and federal agencies on environmental issues.

Dr. Reinhard Pienitz (reinhard.pienitz@cen.ulaval.ca) is a Professor in the Department of Geography at Université Laval (Québec City, Canada), with an adjunct professorship at Carleton University (Ottawa, Canada). He directs the Paleolimnology-Paleoecology Laboratory at the Centre d'Études Nordiques (CEN). He presently supervises and co-supervises the research of 15 graduate students and post-docs in the fields of paleolimnology and paleoceanography, in both the departments of geography and biology. The interdisciplinary research in his laboratory focuses on the use of modern and fossil algae and insects as indicators of environmental change in lakes and rivers of Arctic and temperate regions, as well as of marine coastal regions. His laboratory has administered the Circumpolar Diatom Database (CDD) since 1997.

Dr. Tom Murphy (Tom.Murphy@ec.gc.ca) is a Research Scientist at the National Water Research Institute in Burlington, Canada. He received the following degrees: B.Sc., Biology, Queens University, Canada; M.Sc., Botany, University of Toronto, Canada; Ph.D., Zoology, University of British Columbia, Canada; Sc.D., State University of New York, USA. His expertise includes assessment and control of lake eutrophication and pollution including in situ bioremediation of toxic sediments and in-lake reduction of eutrophication.

Dr. Hans Peterson (hanspeterson@sasktel.net) is President of WateResearch Corp., which is an R&D company working on innovative solutions to wastewater remediation and drinking water treatment. In the past couple of years WateResearch Corp. has carried out contract research for the U.S. Environmental Protection Agency, Danish International Development Agency, Health Canada, Environment Canada, several municipalities and private industry. WateResearch Corp. is currently isolating algae for the removal of contaminants from waste lagoons and microbial consortia for the removal of iron, manganese, arsenic, and ammonium from drinking water. Hans is also the Executive Director of the Safe Drinking Water Foundation, which promotes international collaboration on drinking water issues in rural areas (http://www.safewater.org).

Dr. Louis Legendre (legendre@obs-vlfr.fr) is a Research Professor at the Centre National de la Recherche Scientifique, France, and Emeritus Professor at Laval University, Canada. He is Director of the Villefranche Oceanography Laboratory, Villefranche-sur-Mer, France. Professor Legendre is a specialist of biological oceanography and numerical ecology, with special interest in the regulation exerted by pelagic ecosystems on the biogeochemical fluxes of carbon in aquatic environments.

Dr. Masumi Yamamuro (http://staff.aist.go.jp/m-yamamuro/) is a Senior Researcher at the Institute for Marine Resources and Environment, Geological Survey of Japan, AIST. She has been studying benthic ecosystems in coastal areas including estuarine lagoons, seagrass beds, and coral reefs. One of her contributions to ecology is suggesting the importance of nitrogen fixation in coral reef ecosystems based on her analytical results of nitrogen stable isotope ratios. Dr. Yamamuro is a council member of both the Oceanographic Society of Japan and the Japan Society of Endocrine Disrupters Research. She was awarded the 2000 Biwako Prize for Ecology.

Dr. Erik Jeppesen (ej@DMU.dk) is a Research Professor in a joint position at the National Environmental Research Institute, Silkeborg and the University of Aarhus, Denmark. He is a specialist in shallow lakes ecology and restoration. His main interests are trophic dynamics and nutrient cycling in lakes along gradients in climate (from the Arctic to the subtropics) and nutrients, palaeoecology and ecosystem modeling. He is the founder of the continuing international conference on 'Shallow Lakes Ecology' and is on the editorial advisory board of the *Journal of Aquatic Ecosystem Health & Management*, as well as on the editorial board of the journal *Ecosystems*. In 1998 he became a Doctor of Science with the thesis 'The ecology of shallow lakes - trophic interactions in the pelagial.'

Dr. Clive Howard-Williams (c.howard-williams@niwa.co.nz) is General Manager (Freshwater) of New Zealand's National Institute of Water and Atmospheric Research, and a Fellow of the Royal Society of New Zealand. He has had experience in limnology in many parts of the world from the tropics to the polar regions. He is currently on the editorial boards of *Wetlands Ecology and Management*, and *Antarctic Science*. He has led research programmes on lake, river, and wetland ecosystems, and on the impacts of invasive plant species on aquatic systems. In his present position he co-ordinates both research and environmental consultancy work, and has a special interest in the human dimensions relating to environmental work, particularly those of involving communities and industry in the results of research.

Dr. Charles R. Goldman (crgoldman@ucdavis.edu) is Professor of Limnology in the Department of Environmental Science and Policy at the University of California Davis. He has directed the Lake Tahoe Research Group since 1959. He has specialized in nutrient limiting factors for algal growth and the eutrophication of lakes with emphasis on high-altitude and high-latitude lakes throughout the world. He has trained over 100 graduate students and postdoctoral associates during his 43-year career at the University of California and recently received the first graduate student mentoring award from the University for this record. He has published more than 400 journal articles and books. His research has included tropical and temperate lakes from the Arctic to the Antarctic where the Goldman glacier was named in his honor. His best-known work is the long-term research programme at Tahoe and Castle Lakes. In July of 1997 he hosted President Clinton and Vice President Gore aboard the University Research Vessel *John le Conte* and in 1998 received the Albert Einstein World Prize in Science. In 2003 he received the Nevada Medal of Science for his career-long efforts to preserve the water quality of Lake Tahoe.

Dr. Mark James (m.james@niwa.co.nz) is an aquatic ecologist and a Director of the National Institute of Water and Atmospheric Research in New Zealand. He has over 20 years experience working in lakes in New Zealand, Antarctica, Denmark, and Finland and has published extensively on aquatic ecosystems. He has led large multidisciplinary lake and coastal research programmes and for the last 10 years has been involved in a large number of projects aimed at sustainable lake management. He has worked closely with local bodies and communities, industry, and the scientific community to ensure an integrated approach involving lake and catchment management and has been heavily involved in recent initiatives to manage nutrient inputs to lakes to prevent further deterioration.

Dr. Warwick F. Vincent (warwick.vincent@bio.ulaval.ca) is Professor of Limnology and Canada Research Chair in Aquatic Ecosystem Studies at Laval University, Québec City, Canada. He has worked on a wide range of freshwater and marine ecosystems throughout the world, including Lake Taupo (New Zealand), Lake Titicaca (Peru-Bolivia), Lake Tahoe (USA), Wastwater (England), Lake Biwa (Japan), and the St. Lawrence River (Canada-USA). His research team has a special interest in cyanobacteria and microbial processes at the base of aquatic food webs, and much of their current research concerns the effects of climate on high-latitude lakes, wetlands, rivers, and coastal seas. He is a Fellow of the Royal Society of Canada and an Honorary Fellow of the Royal Society of New Zealand.

Chapter 1

Lessons from Lake Biwa and Other Asian Lakes: Global and Local Perspectives

Michio Kumagai[1], Warwick F. Vincent[2], Kanako Ishikawa[1] and Yasuaki Aota[1]

[1] *Lake Biwa Research Institute, 1-10 Uchidehama, Otsu, 520-0806, Japan*

[2] *Dépt de Biologie, Université Laval, Québec, QC G1K 7P4, Canada*

Abstract

The freshwater resources of Asia are under enormous pressure given the high and increasing population densities of the region, and the serious degradation of many lakes and rivers due to agricultural, industrial and urban development. Global climate change is likely to exacerbate these pressures through changes in the hydrological balance; influences on stratification, algal blooms and deep water re-oxygenation; and for high altitude sites such as lakes on the Mongolian Plateau, accelerated permafrost melting and desertification. Eutrophication and other processes of water quality degradation are especially severe in many parts of China. At Lake Taihu, a drinking water supply for 40 million people in central China, toxic cyanobacterial blooms now occur throughout the year. In the Province of Yunnan, many of the shallow lake waters are highly polluted and are of human health concern. Lake Biwa, Japan's largest freshwater body, has experienced many environmental problems over the last 50 years including loss of species habitat, changes in oxygen content, and blooms of noxious cyanobacteria. Two multi-disciplinary research programs involving specialists in hydrodynamics, bio-optics, biogeochemistry and freshwater ecology have generated new insights into the structure and functioning of the Lake Biwa ecosystem, and have contributed an improved understanding of the processes affecting water quality. The Biwako Transport

Experiment (BITEX'93) provided new information about physical-biotic coupling in the lake, and revealed the stimulatory effects of typhoons on south basin populations of phytoplankton. Measurements of water currents also showed that toxic cyanobacteria can be transported via a reverse surface flow from the South to North Basins during typhoon events. A subsequent program, Cyanobacterial Risk Assessment at Biwako (CRAB), resulted in predictive bio-optical models and elucidated the importance of stratification and advection processes for cyanobacterial bloom development. The algal bloom populations are favored by diurnally stratified, nutrient-rich conditions inshore and are then advected via a gyre into the main basin of the lake where the cells can continue their growth on stored reserves. The successful management of Lake Biwa, and lakes elsewhere, will especially depend on an improved scientific understanding of the limnological controls on water quality. It will also be fostered by cross-disciplinary and international exchanges of information, and a commitment by scientists, residents and others with a vested interest in water quality, including regulatory authorities, to work together towards the common goal of long term, integrated protection of the lake and its surrounding watershed.

Introduction

The availability of safe drinking water, sustained water supplies for agriculture, and the protection of freshwater habitats for aquatic wildlife are major pre-occupations throughout Asia. This region contains most of the world's population, and continues to experience rapid population and economic growth. Many of the aquatic ecosystems in Asia have been severely impacted by local agricultural, industrial and urban development, and their rehabilitation and long term protection are now key priorities for sustained economic growth and well-being. New concerns are also emerging, especially the potential impact of global warming trends on water quantity as well as quality.

In this chapter we present examples of aquatic research in Asia related to the theme of global versus local issues in freshwater management. Firstly, we examine the influence and potential implications of global climate trends on deep lakes in Asia, and some of the water quality issues facing lake managers in Mongolia, China and Japan. We then focus on Lake Biwa, Japan, and describe some of the insights derived from a multi-disciplinary, multi-national research strategy towards improved limnological understanding and stewardship of this key freshwater resource.

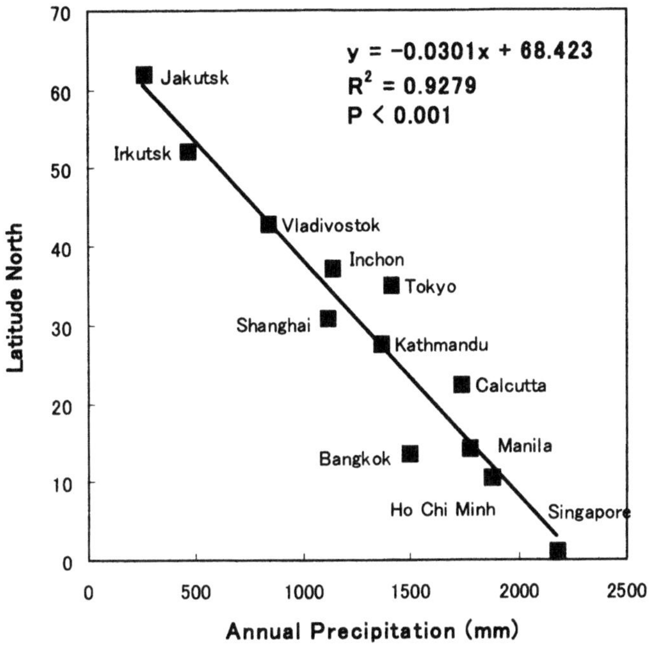

Fig. 1. Relationship between annual precipitation and latitude north in Asia.

Global and Local Issues for Water Management in Asia

Climate change in Asia is of pressing environmental concern for the global as well as regional economy and society. This area has been blessed with a relatively stable climate until now, and may now be entering a period of major instability as a result of global change. Figure 1 shows the relationship between annual precipitation and latitude north in the Asian main cities (Nippon Astronomical Observatory, 2002). The remarkably high correlation coefficient for this relationship ($r = -0.96$, $N=12$, $P<0.001$) reflects the regulating effects of the vast Eurasian Continent and Pacific Ocean. The sizes of these two components of the planetary system mean that they have not only regional effects on climatic stability, but also exert an influence at a global scale.

The natural stability of the regional climate has brought considerable benefits to Asia. Bountiful rain in China, India and some South-East Asian countries has enabled the cultivation of rice that in turn has supported a large fraction of the world's human population. Rice has a high productivity and requires much water, and therefore a predictably abundant water supply has been essential. Future shifts in hydrological

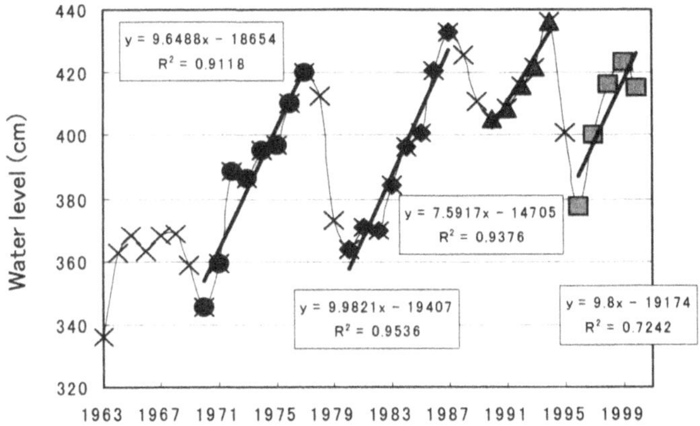

Fig. 2. Temporal change of water level at Lake Hövsgöl

balance associated with climate change would have serious implications for the rice-based agricultural systems throughout this region. Other parts of Asia experience more limited rainfall. For example, the Mongolian Plateau receives relatively low precipitation but the temperature is moderate in summer and has allowed the production of high quality grass to keep livestock such as sheep, goats and cows. Throughout Asia, the balance of precipitation and temperature is critical to the primary production that sustains agriculture and the human populations (Lieth, 1975).

There is now evidence of climate-related impacts on land and freshwater ecosystems in parts of Asia. In highland areas such as Tibet and Mongolia, desertification is expanding, and the nomadic people are forced to move to other sites for water. The hydrology of lakes and rivers appears also to be changing.

The first evidence of potential climate impacts on lake ecosystems is from monitoring records of air temperature. The air temperature near Lake Hövsgöl (50°30′-51°35′N, 100°15′- 100°40′E) in Mongolia increased at $0.13℃yr^{-1}$ ($r=0.86$, $N=20$, $p<0.001$) over the period 1981 to 2000. Concomitant with this air temperature rise, Lake Hövsgöl showed several multi-year periods of rise in water level punctuated by abrupt drops (Fig. 2). For the period 1970 to 2000 there was an overall increase of $9.23\pm1.11cm\ yr^{-1}$ that may be related to increased glacier-melt or permafrost-melt due to the climatic warming (Kumagai et al. 2003). The river mouth of the Eguiin Gol, the only outlet from Lake Hövsgöl, was sometimes buried with debris after heavy rains, contributing to the high water levels. Whenever the water level has become too high, the village government has excavated sand and stones from the river to increase outflow and rapidly lower the lake. Any long term loss of permafrost will bring serious impacts not only on the lakes, but also on the availability of water in the surrounding terrestrial ecosystems. Large scale permafrost melting will reduce and eliminate the major reservoir of water that becomes partially available and is recharged each year. This plus increased evaporation, and the resultant desertification, will have severe impacts on the cattle farming practices that support the nomadic people of this region (Kumagai and Urabe, 2002).

Water is an especially critical and complex issue for China. The northwest part of China is experiencing loss of water supply, and the combined effects of over-irrigation and frequent droughts have severely affected water use in the capital city Beijing (Brown and Halweil, 1998). Since 1997, it is reported that the Great Yellow River has been completely dry near its mouth for several months each year. The National Government in China is now planning to harness water from the south, specifically from water-rich areas along the Yangtze River, which experiences annual flooding due to heavy rain. In the central region of China near Shanghai and Nanjing, eutrophication is the major issue because so many people flush their wastes directly into lakes and rivers without any treatment. Also many industrial regions are expanding in the area between Nanjing and Shanghai. Lake Taihu (30°56′-31°34′N, 119°54′- 120°36′E), a vast lake that provides drinking water to 40 million people of this central China region, is subject to steadily worsening eutrophication (Pu et al. 1998). Twenty years ago, noxious cyanobacteria (blue-green algae) formed blooms that were restricted to limited areas and only during summer. Such blooms now occur throughout the lake in all seasons. Even boiled tap water has an unpleasant taste and odour and has resulted in increased demand for bottled drinking water. Recent flooding has also compounded the environmental problems in this region, and underscores the difficulties that environmental managers may need to face if ongoing climate trends cause large, unpredictable changes in weather.

Eutrophication is also a serious issue for Yunnan Province located in the south of China on the highland plateau (1000 to 3000 m altitude) facing Vietnam, Laos and Myanmar. Most of the lakes in this province have suffered nutrient enrichment and other pollutants due to heavy industrial waste non-point sources including agriculture and urban wastes. Lake Dianchi (24°22′-26°33′N, 102°10′- 103°40′E) near Kunming has especially degraded water and other shallow lakes nearby have similarly poor quality environments. There is currently a proposal to build an International Research Center for Yunnan Plateau Lakes that would bring together international expertise in

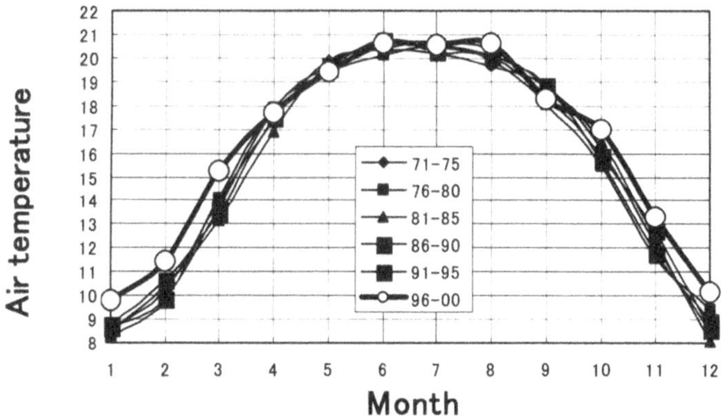

Fig. 3. Comparison of average air temperature over 5 years at Lake Fuxian from 1971 to 2000.

water quality research and management to restore and protect these waters.

Global change may exacerbate the water quality effects of local pollution. Changes in the water balance can lead to prolonged hydraulic residence times, in turn favouring higher total phosphorus, pollutant concentrations and algal biomass. Increased temperatures could lead to stronger cycles of diurnal stratification, in turn encouraging cyanobacterial blooms. Such effects may also influence the seasonal stratification regime leading to problems that are especially severe in deep lakes. Small increases in air temperature can result in a reduced extent of convective and wind-induced mixing with the result that deep lakes fail to mix completely in winter, have less recharge of oxygen and a longer period of isolation, ultimately leading to bottom water anoxia. For example, Lake Fuxian (24°21′-24°38′N, 102°49′- 102°57′E), the second deepest lake in China (maximum depth = 155m, area = 211km^2 and volume = 18.9km^3 ; NIGLAS, 1990), has recently experienced increased air temperatures. Specifically, average air temperature during the period of winter mixing between 1996 and 2000 was nearly 1 ℃ higher than for the preceding 5-years (Fig. 3). A reduced extent of winter mixing was observed, and increased deep water depletion of oxygen.

These impacts of year-to-year variations in climate have been observed elsewhere in the world. For example, Lake Titicaca (Peru-Bolivia) varies greatly between years in its degree of mixing, extent of deepwater anoxia and resultant accumulation or loss of nutrients (Vincent et al. 1985). There is some evidence of a trend towards increased

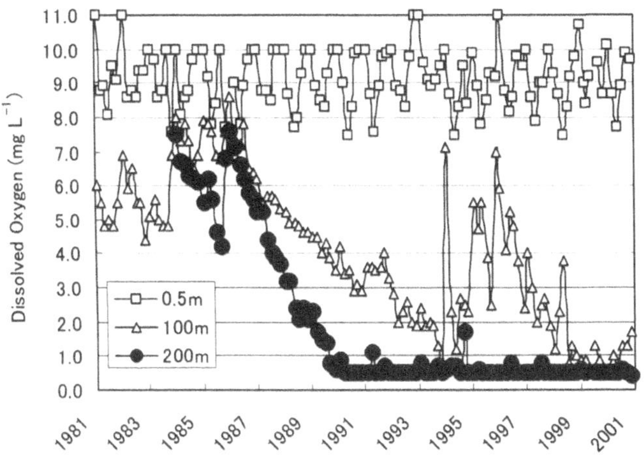

Fig. 4. Temporal change of dissolved oxygen at Lake Ikeda. Symbol □ shows the data at 0.5m depth, △ 100m depth and ● 200m depth.

bottom water oxygen depletion at Lake Geneva influenced by climate warming (Ishiguro, 2002). One of the most striking examples is from Lake Ikeda (31°14′N, 130°34′E) in the Kagoshima Prefecture of Japan. This lake is deep (maximum depth = 233 m) but relatively small (area = 10.95km^2 ; volume = 1.47km^3). The temporal changes of dissolved oxygen concentration at three different levels of 0.5m, 100m and 200m depths are depicted in Figure 4. The dissolved oxygen at 200m depth started to decrease in 1984 and recovered in 1986. However it decreased again in 1987, became almost zero in 1990, and has never recovered subsequently. Dissolved oxygen concentrations at 100m depth gradually decreased from 1987 to 1993, but suddenly recovered in 1994. From 1994 to 1998, dissolved oxygen at 100m depth fluctuated between 1 and 7 mg L^{-1}, and then became zero again. These results illustrate how easily the precarious balance of oxygen gains and losses can be perturbed by climate, and also show that bottom water anoxia can persist for long periods of time and may be difficult to reset to fully oxygenated conditions. This balance is sensitive to small changes in air temperature and rainfall in winter. Cooled surface waters bring oxygen to the hypolimnion in deep lakes through vertical convection and/or turbulent diffusion processes (Hosoda and Hosomi, 2002), while cold, dense river water such as snowmelt can intrude directly along lake bottom as a density current rich in oxygen (Kumagai and

Fushimi, 1995). The latter recharge of bottom water oxygen by inflowing density currents in winter may be an important factor reducing the effects of eutrophication in many deep lakes, for example Lake Rotoiti, New Zealand (Vincent et al. 1991).

Historical Changes at Lake Biwa

Lake Biwa (34°58′-35°31′N, 135°52′- 136°17′E) is Japan's largest freshwater lake and its limnological properties, water quality management and environmental protection have therefore received much attention. It is an ancient lake with a nearly 5 million year history. The original basin was positioned some 100 km south of its present location, and it has moved northwards by tectonic processes. The lake reached its current position 400,000 years ago, and continues to move to the north and sink at several mm per year (Horie, 1984). The name Biwa appears to be derived from its shape which resembles the ancient mandolin-like instrument of the same name, although it is unsure as to whether the early lake-dwellers would have been able to see the entire shape from the mountains around the lake (Kimura, 2001). Lake Biwa consists of two basins: a large and deep North Basin with an area of 612 km^2, maximum depth of 104 m and mean depth of 44 m; and a small and shallow South Basin with an area of 58 km^2, maximum depth of 8 m and mean depth of 3.5m. The total volume of the lake is 27.5 km^3, and the residence time is 5.5 years (Kira, 1984). One of the most remarkable features of Lake Biwa is that the portions of the North Basin deeper than 85 m are below the mean sea level at Osaka Bay. This considerable depth also indicates its sensitivity to change and the potential difficulties of lake rehabilitation.

The mean annual precipitation over Japan has fallen from 636 km^3 between 1887 and 1906 to 607 km^3 between 1987 and 1997. Thus Japan has lost about 29 km^3 (4.4%) of rainfall for last 100 years, almost same volume as Lake Biwa. As the total volume of freshwater lakes in Japan is close to 84 km^3, nearly 14% of fresh water supplied by rainfall can be saved in the lakes. The total discharge from the rivers in Japan is almost 250 km^3, accounting for 41% of total precipitation. Another 45% of fresh water may be stored in soil, forest, and ground water (Nippon Astronomical Observatory, 2002). The loss of 4.4% fresh water over the course of the last century is significant and likely due to decreases of winter snowfall as well as reduced precipitation during the typhoon season in late summer. The reduced input of cold snowmelt may have a particularly strong influence on hypolimnetic oxygen (as discussed above), which has been steadily declining. As seen in Figure 5, the absolute difference between mean air temperature and 4 ℃ during January and March is inversely related to the minimum dissolved

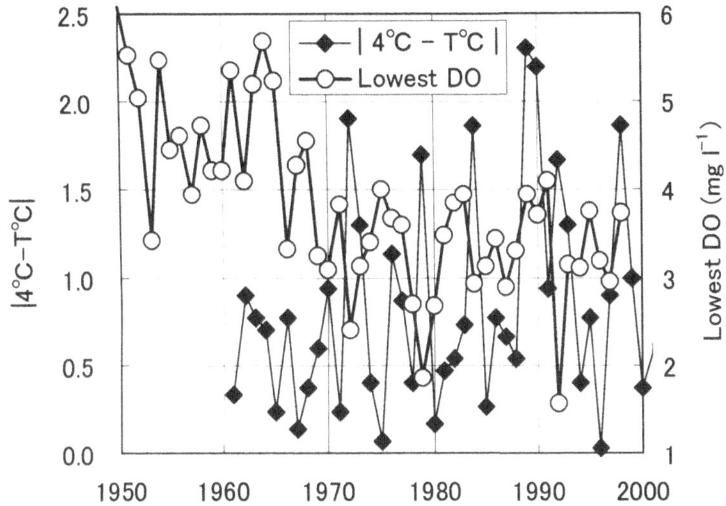

Fig. 5. Absolute difference between mean air temperature and 4℃ during January and March measured by Hikone Meteorological Observatory (◆), and the lowest dissolved oxygen of every year measured by Shiga Prefectural Fisheries Experimental Station (○).

oxygen concentrations in the bottoms water (70 m) of the North Basin. Gradual eutrophication has also been a major contributor to the persistent decline in deep water oxygen in the lake.

Lake Biwa has a rich biodiversity, with 500 plant species (including algae), 600 animal species, including 9% endemic taxa that reflect the long history of the lake (Mori and Miura, 1990). The benthic aquatic communities of Lake Biwa changed dramatically between 1963 and 2000, with the greatest shifts in 1970 and 1987. One of the endemic fish species of Lake Biwa, *Chaenogobius isaza* Tanaka (common name = Isaza), resides in waters deeper than 30 m through most of the year, coming into surface waters from April to June for spawning in the shallow inshore area. Nakanishi and Nagoshi (1984) have shown that a substantial change in the food habits of Isaza occurred after 1970, as indicated by an examination of stomach contents. Before 1970, Isaza mainly fed on zooplankton, but it changed prey and started to feed on the amphipod *Jesogammarus annandalei* after 1970. Nakanishi and Nagoshi (1984) attributed this change to eutrophication, although competition for resources by the planktivore. *Plecoglossus altivelis altivelis* Temminck et Schlegel (common name = Ayu) may also have played a role. The catch of Ayu between 1960 and 1968 was 3,687(sd=1,637) ton yr^{-1}, but greatly increased to 9,247(sd=2,056) ton yr^{-1} from 1969 to 1979, and further to

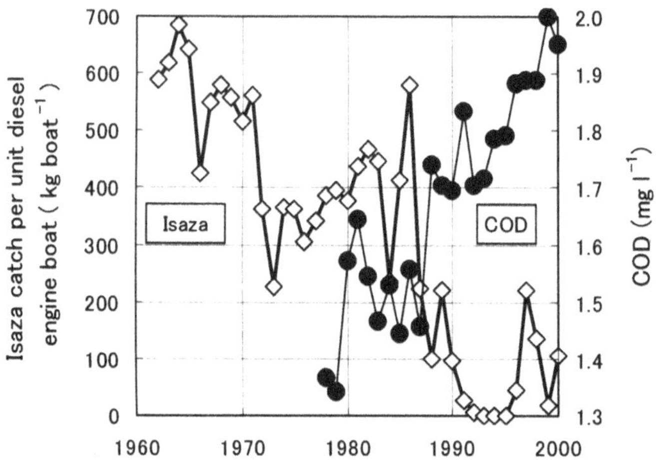

Fig. 6. Temporal changes of Isaza (*Chaenogobius isaza* TANAKA) measured by Shiga Prefectural Fisheries Experimental Station (◇) and COD measured by Shiga Prefectural Institute of Public Health and Environmental Science (●) in Lake Biwa.

15,252(sd=3,764) ton yr^{-1}, over the period 1980-1992. These increases were associated with culture and release of this species from fish hatcheries. Like Isaza, Ayu also feeds on zooplankton, and competition by the higher population densities of this species may have contributed towards the shift in diet by the Isaza. These observations indicate the importance of understanding local biological conditions for long term lake management.

As seen in Figure 6, the Isaza catch suddenly dropped after 1987, at the same time that the chemical oxygen demand (COD) below 80 m depth started to increase. There was a significant inverse relationship between the two variables: Isaza catch per unit boat = -847.43 x (COD) + 1623.6, R^2=0.82, N=22, $p< 0.001$). More detailed analyses, however, suggested that the decrease of Isaza catch occurred one year prior to the decrease of COD. There also appear to be cascading foodweb effects. After the Isaza declined in 1987, the density of *J. annandalei* rapidly increased (T. Ishikawa, pers. comm.). The abundance of *J. annandalei* varied between 200 and 63000 m^{-2} in 1997 and 1998, and its biomass (0.01 to 3.6 g C m^{-2}) was almost comparable with that of zooplankton (Ishikawa and Urabe, 2002). Although cause and effect relationships for the precipitous decline of Isaza have yet to be clarified, the deterioration of the deep waters is likely to have had a serious influence on this species.

Ogawa et al. (2001) analyzed nitrogen isotope ratios of Isaza and sediments in Lake Biwa, and concluded that the increase of $\delta^{15}N$ after 1968 was caused by eutrophication and denitrification. There is a strong positive correlation between the concentration of persistent organic pollutants in the sediments, specifically total PCDDs (Sakai et al. 1999) and the $\delta^{15}N$ signature of the Isaza (R^2=0.88, N=8, $P<0.001$). The organic pollutants PCDDs and PCDFs have been continuously increasing since 1960, despite the cessation of PCP(pentachlorophenol) production in the 1960s and of CNP(chlornitrofen) production in 1970s. The results suggest that some PCDDs/DFs are still entering the lake as dust from garbage burning or from flushing of dioxin residues in the soil. The similar accumulation of PCDDs/DFs in the sediments can be seen in Lake Shinji (Masunaga et al. 2001). The accumulation of PCDDs and $\delta^{15}N$ in the sediments is also consistent with increasingly anoxic sediment conditions, and has occurred during a period of gradually depleting dissolved oxygen in the hypolimnion of Lake Biwa since 1968 (Naka, 1973). Recently, filamentous sulfur-oxidizing bacteria have been found in Lake Biwa, and are another indicator of anoxic sediments (Nishino et al. 1998). Decreased oxygen availability may slow the decomposition rate of PCDDs in lake sediments, leading to enhanced accumulation. Similarly, denitrification is likely to be enhanced by anoxia, resulting in the high $\delta^{15}N$ values (Yamada et al. 1996). The transfer of these toxin-contaminated sediments into the water column, for example by wind-induced resuspension or overturning, will require close attention in the future water management of water quality in Lake Biwa.

Insights From the Biwako Transport Experiment (BITEX'93)

The Lake Biwa Research Institute and the University of Western Australia jointly organized an international field experiment in Lake Biwa named BITEX'93 (**Bi**wa-ko **T**ransport **Ex**periment). Some 177 researchers, technicians and students, representing seven countries, gathered at the lake and worked together for almost one month in the summer of 1993. It was one of the largest field experiments in limnology, and focused on biotic-physical interactions, specifically the biogeochemical responses of the Lake Biwa ecosystem to typhoon-induced mixing. A core program of basic field measurements was conducted (Kumagai and Nakano, 1996), including background chemical data measured every two days during the experiment by Hashitani et al. (1996) and Seike et al. (1996). A broad range of additional observations and experiments were undertaken against this backdrop of continuous limnological data.

Three typhoon events occurred during the experiment and gave rise to internal waves and surges that mixed the lake (Jiao and Kumagai, 1995; Hayami et al. 1996; Saggio and Imberger, 1998). Toda (1996) used NOAA-AVHRR satellite data to monitor surface water temperature, and Melack et al. (1996) obtained optimum linear regressions, which can be adapted to the airborne remote sensing data. These data provided broader synoptic coverage but compared well with field data taken by Tanaka and Tsuda (1996). The strong typhoon winds mixed both the North and South Basins of the lake, but the reaction in each basin was quite different. MacIntyre (1996) calculated overturning scales under different weather conditions. In the shallow South Basin, typhoon mixing caused a resuspension of bottom sediments and entrained nutrients into the water column (Sakamoto and Inoue, 1996; Robarts et al. 1998). The latter enhanced the growth of *Aulacoseira granulata* (Nakano et al. 1996) and primary production (Frenette et al. 1996a; Nagata et al. 1996), but did not significantly affect bacterial production because there was little transfer of labile DOM (Ogawa and Nagata, 1996). No influence of typhoon mixing on the abundance of planktonic crustacea was observed (Kawabata and Urabe, 1996), but there were potentially effects on their grazing rates (Urabe and Kawabata, 1996).

Trevorrow (1996) measured near-surface air bubbles with six vertical sonar (Sea Scan, 29 to 397 kHz), and found the bubble penetration depth of 1.8 to 2.2 m with 10 to 16 mm radius under winds of 10 m s^{-1}. This technology was also adapted to monitor the population density of the amphipod populations of *J. annandalei* (Trevorrow and Tanaka, 1997), and demonstrated its utility for detecting the nocturnal migration of this species. Material transport originating from discharge of rivers with high turbidity, and due to resuspension of bottom sediments by internal wave-breaking, was also measured by Tanaka et al. (1996) and Kumagai et al. (1996), respectively.

Vertical mixing due to the typhoons reduced the water column stability and influenced the size structure of the phytoplankton community. There were differences in the responses of picoplankton and nanoplankton to mixing in the shallow South Basin (Frenette et al. 1996a) and the deep North Basin (Frenette et al. 1996b), reflecting differences in strategies of photoadaptation between the two size fractions (Frenette et al. 1998). The results from the North Basin also underscored how the surface layer of the lake (epilimnion) was not homogeneous but consisted of a sandwich of layers that differed in algal physiological characteristics (Frenette et al. 1996b) and mixing properties (Robarts et al. 1998).

The central hypothesis of the BITEX experiment was that lake water as well as its dissolved and suspended materials would be effectively exchanged between the North Basin and the South Basin by Kelvin wave sloshing induced by the typhoons.

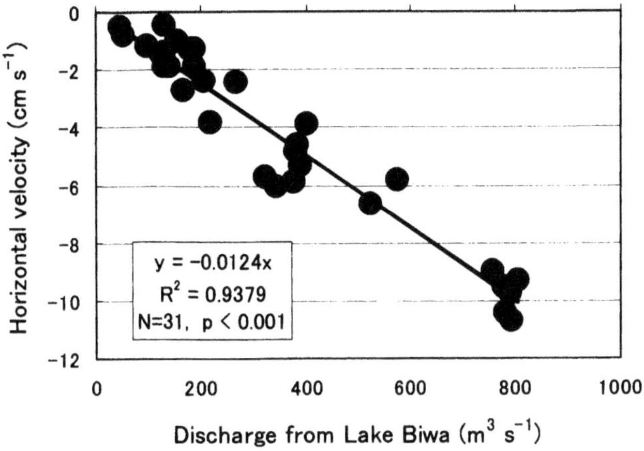

Fig. 7. Relationship between discharge from Lake Biwa and southward horizontal velocity at Lake Biwa Bridge.

Contrary to this hypothesis, this sloshing effect was not observed and the two basins seemed to respond independently to the typhoon events (Frenette et al. 1996c; Nagata et al. 1996). The North Basin microbial food web responded slowly while there was a rapid and strong response by the South Basin biota in response to sediment resuspension. In general, the mean horizontal southward velocities measured at the Lake Biwa Bridge (between the two basins) is strongly and positively correlated with the discharge from Lake Biwa, measured as the sum of water discharge from the Seta River, Uji water power station and Kyoto drinking water gate (Fig. 7). This, however, represents mean velocity of four current meters, and some of them measured northward currents induced by strong winds such as the typhoon. These inverse currents have the potential to transport some algal species from the eutrophic South Basin to the mesotrophic North Basin. As seen in Figure 8, during BITEX'93, there was a strong inverse flux of 20–90 m^3 s^{-1} from the South to North Basin on September 2 and 4, 1993 after a large typhoon (T9313) passed. The cell density of *Microcystis wesenbergii* at St.B, which was located at 35°06′56"N, 135°56′14"E near the Lake Biwa Bridge, peaked on September 2, and there was a similar peak of *M. wesenbergii* at St.N (35°10′36" N, 136°58′01" E) in the North Basin on September 5. These observations strongly imply a significant biological exchange between the two basins, and a mechanism whereby South Basin populations could seed blooms that subsequently develop in the North Basin. Thus typhoon events could be influential in affecting water quality properties throughout the lake.

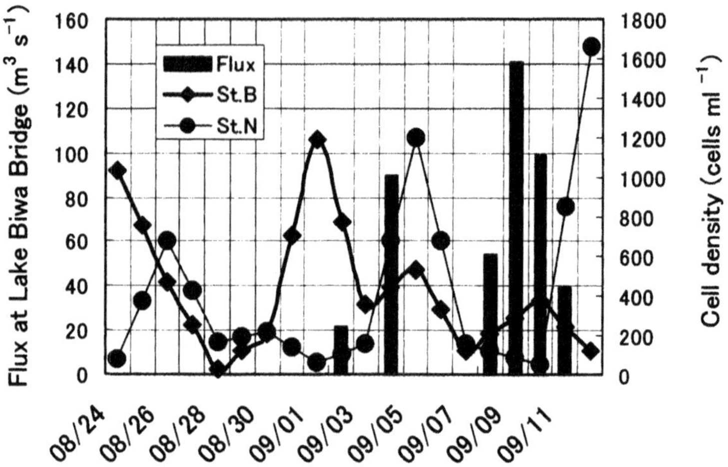

Fig. 8. Cell densities of *Microcystis wesenbergii* at St.B and St.N, and the northward flux at Lake Biwa Bridge.

Cyanobacteria Risk Assessment in Biwako (CRAB)

Subsequent to BITEX'93, an international program to assess the factors controlling the growth of toxic cyanobacteria in Lake Biwa was initiated. This program, the **C**yanobacteria **R**isk **A**ssessment for **B**iwako, included detailed, coupled observations of the physical, chemical and biological properties across the lake, as well as a series of large-scale enclosure experiments at Akanoi Bay, a large eutrophic bay in the South Basin.

The summers of 1994 and 1995 were distinguished by very small amounts of rainfall, and cyanobacterial blooms occurred many places in Lake Biwa. Blue-green algal scums were conspicuous in Akanoi Bay, and created substantial public anxiety about the magnitude of deterioration in the water quality of Lake Biwa. The dominant species of cyanobacteria found in Akanoi Bay in 1995 were *Anabaena affinis*, *Microcystis aeruginosa* and *Planktothrix agardhii*, with differences between taxa in their distribution attributed to different responses to wind driven currents (Ishikawa et al. 1999). In 1994, *Microcystis sp.* blooms occurred not only at Akanoi Bay but also Nagahama Harbour in the North Basin, and within 2-3 years cyanobacterial problems were apparent throughout the North Basin. The enclosure experiments were undertaken to examine the growth and development mechanisms of cyanobacterial blooms. Among the many potential controls on water blooms, temperature is a key variable influencing both directly and indirectly cyanobacterial growth and dominance (Vincent, 1989). Both

Microcystis sp. and *Anabaena sp.* cultured from Akanoi Bay had maximum growth rates at 32℃ (Nalewajko and Murphy, 2001), and once they developed *in situ*, they absorbed heat and thereby enhanced the temperature stratification (Kumagai et al. 2000). Diurnal stratification was also one of key factors for cyanobacterial blooms, because *Microcystis sp.* and *Anabaena sp.* are able to migrate down at night and up at daytime in stratified waters (Nakano et al. 2001). Hayakawa et al. (2002) measured the fatty acid composition of *Microcystis aeruginosa*, and showed that it was related to growth rate.

The sediments play a variety of roles in the growth of cyanobacteria, firstly via nutrient release and secondly as a refuge and 'seed bank' of the resting stages of cyanobacteria, such as the akinetes of *Anabaena*. Sediment studies at the enclosure site and the North Basin provided insights into the biogeochemical mechanisms of phosphorus release, and showed the opportunities for chemical treatment of enriched sediments as a strategy towards suppressing cyanobacterial blooms (Murphy et al. 2001; Murphy et al. 2002; Murphy and Kumagai 2002).

Following the termination of the enclosure experiments at Akanoi Bay in 1997, we examined the food web and distributional ecology of cyanobacteria in Lake Biwa. Elser et al. (2001) found that zooplankton grazing inshore may reduce algal blooms due to direct feeding, but also has the potential to change the ratios of N:P supply that affect cyanobacterial growth. By applying the carbon steady-state model of the planktonic food web to Lake Biwa, Niquil et al. (unpublished) showed that the microbial system of the planktonic community is very active, and the non-living organic carbon is directly consumed by mesozooplankton. Light is also an important controlling factor for cyanobacterial blooms. UV attenuation in Lake Biwa was measured by Vincent et al. (2001) who found that the penetration of these damaging wavelengths is affected by particulate material as well as by dissolved organic matter. This work was then extended at Lake Biwa by Belzile et al. (2002). They developed a general model of UV and PAR attenuation in natural waters that incorporates both scattering as well as absorption. This approach has been subsequently applied to the water quality issues of lake water colour and transparency at Lake Taupo, New Zealand (Belzile et al. unpublished).

Many aspects of the life cycle of bloom-forming cyanobacteria have yet to be fully understood, and one of the major unknowns is the relationship between water column and sediment populations. Tsujimura et al. (2000) studied the seasonal variation of *Microcystis sp.* populations in Lake Biwa sediments. Experimental studies showed that once *Microcystis sp.* sinks below the euphotic zone, it cannot be recruited again because of insufficient light (Ishikawa et al. in press).

As seen in Figure 9, *Microcystis aeruginosa* is now thoroughly distributed

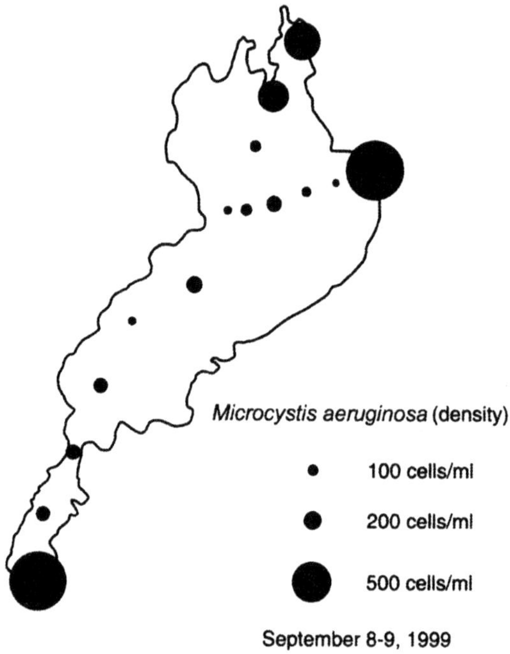

Fig. 9. Horizontal distribution of *Microcystis aeruginosa* measured in Lake Biwa on September 8-9 in 1999.

throughout the North Basin of the lake (Kumagai et al. 2001a). Part of the research program has therefore focused on where the seed populations may come from in the North Basin. We have investigated the surface water currents in the North Basin with ADCP (Acoustic Doppler Currents Profiler) since 1993. Three gyres are formed in Lake Biwa during the period of stratification from May to October, and in particular the First Gyre is big and stable enough to transport material from inshore to offshore (Kumagai et al. 1998). Ishikawa et al. (2002a) showed a spatial relationship between *Microcystis sp.* and a gyral current in the lake (the First Gyre of Lake Biwa). However the populations were unable to grow in the low nutrient waters at the center of gyre. These populations appear to begin inshore and are then advected offshore where they continue to grow on stored reserves (Ishikawa et al. 2002b).

Future of Lake Biwa Management

Our research on Lake Biwa has revealed a variety of changes that affect its water quality. Many of these changes will be difficult and slow to reverse. Lake Biwa is a deep lake with a long retention time, and therefore the potential for local accumulation of nutrients, contaminants and undesirable biota, such as spores of toxic cyanobacteria, is substantial. Reversing and flushing out the effects of this accumulation will require a long term commitment. Ongoing changes in climate may exacerbate these effects by reducing the deep transport of oxygen-rich inflows, and the extent of winter mixing. All of these observations imply that the monitoring of Lake Biwa will require ongoing vigilance, and that emphasis should be placed on first halting further deterioration of the lake and then working towards its improvement relative to current conditions.

We are exploring and developing new technologies to achieve the required level of monitoring. These include a submersible microscope, new bio-optical systems, and autonomous underwater vehicles such as the new AUV 'Tan Tan' developed by LBRI in collaboration with the University of Tokyo, the Ministry of Land, Infrastructure and Transport and the Mitsui Shipbuilding Company (Walker and Kumagai, 2000: Walker et al. 2002: Kumagai et al. 2001b).

The successful management of Lake Biwa, and lakes elsewhere, will depend on an improved understanding of the limnological controls on water quality, cross-disciplinary and international exchanges of information, and a commitment by scientists, residents and other stakeholders including regulatory authorities to work together towards the common goal of long term, integrated protection of the lake and its surrounding watershed. For example, Lake Washington has recovered dramatically from heavy eutrophication after the sewage effluents around the lake were diverted in 1963 to 1967 (Edmondson, 1991). Many of lakes in the world have suffered from similar degradation processes, and improvement to their water quality in the future at a both local and global level will be fostered by the ongoing sharing of information, expertise and widsdom (Kumagai, 2001).

Acknowledgement

We wish to express our sincere gratitude to all participants of BITEX'93 and CRAB programmes. This study was financially supported by the Lake Biwa Research Institute and, in part, by a Grant-in-Aid for Scientific Research (13480161) from the Japan Society for the Promotion of Science.

References

BELZILE, C., W. F. VINCENT, AND M. KUMAGAI. 2002. Contribution of absorption and scattering to the attenuation of UV and photosynthetically available radiation in Lake Biwa. Limnol. Oceanogr. **47:** 95-107.

BROWN, L. R., AND B. HALWEIL. 1998. China's water shortage could shake world food security. World Watch. **July/August:** 10-18.

EDMONDSON, W. T. 1991. The uses of ecology: Lake Washington and beyond. Univ. Washington Press.

ELSER, J., L. GUDEA, M. KYLE, T. ISHIKAWA, AND J. URABE. 2001. Effects of zooplankton on nutrient availability and seston C:N:P stoichiometry in inshore waters of Lake Biwa, Japn. Limnology. **2:** 91-100.

FRENETTE, J-J., W. F. VINCENT, L. LEGENDRE, AND T. NAGATA. 1996a. Size-dependent phytoplankton responses to atmospheric forcing in Lake Biwa. J. Plank. Res. **18:** 371-391.

________, ________, AND ________. 1996b. Size-dependent changes in phytoplankton C and N uptake in the dynamic mixed layer of Lake Biwa. Freshwater Biol. **36:** 221-236.

________, ________, ________, T. NAGATA, K. KAWABATA, AND M. KUMAGAI. 1996c. Biological responses to typhoon-induced mixing in two morphological distinct basins of Lake Biwa. Japn. J. Limnol. **57:** 501-510.

________, ________, ________. 1998. Size-dependent C:N uptake by phytoplankton as a function of irradiance: Ecological implications. Limnol. Oceanogr. **43:** 1362-1368.

HASHITANI, H., M. OKUMURA, AND Y. SEIKE. 1996. Determination of low levels of phosphate phosphorus in environmental waters by preconcentration using activated carbon loaded with zirconium. Japn. J. Limnol. **57:** 479-484.

HAYAKAWA, K., S. TSUJIMURA, G. E. NAPOLITANO, S. NAKANO, M. KUMAGAI, T. NAKAJIMA, AND C. JIAO. 2002. Fatty acid composition as an indicator of physiological condition of the cyanobacterium *Microcystis aeruginosa*. Limnology. **3:** 29-35.

HAYAMI, Y., T. FUJIWARA, AND M. KUMAGAI. 1996. Internal surge in Lake Biwa induced by strong winds of a typhoon. Japn. J. Limnol. **57:** 425-444.

HORIE, S. [ed.]. 1984. Lake Biwa. Dordrecht. Junk.

HOSODA, T., AND T. HOSOMI. 2002. A simplified model to predict seasonal variations of vertical water quality distributions in Lake Biwa and its applications. Advances in River Engineering. JSCE. **8:** 495-500 (in Japanese).

ISHIGURO, N. 2002. Some hypothesis concerning the oxygen supply at the bottom of Lake Geneva. Annals of Ochanomizu Geographical Society (Ochanomizu Chiri). **43:** 33-40 (in Japanese).

ISHIKAWA, T., AND J. URABE. 2002. Population dynamics and production of *Jesogammarus annandalei*, an endemic amphipod, in Lake Biwa, Japan. Freshwater Biol. **47:** 1935-1943.

ISHIKAWA, K., M. KUMAGAI, S. NAKANO, AND H. NAKAHARA. 1999. The influence of wind on the horizontal distribution of bloom-forming cyanobacteria in Akanoi Bay. Japn. J. Limnol. **60:** 531-537.

________, ________, W. F. VINCENT, S. TSUJIMURA, AND H. NAKAHARA. 2002a. Transport and accumulation of bloom-forming cyanobacteria in a large, mid-latitude lake: the gyre-*Microcystis* hypothesis, Limnology. **3:** 87-96.

________, S. TSUJIMURA, W. F. VINCENT, M. KUMAGAI, AND H. NAKAHARA. 2002b. Growth of bloom-forming cyanobacteria under low nutrient conditions. Vech. Internat. Verein. Limnol. **28:** 1-6.

JIAO, C., AND M. KUMAGAI. 1995. Large amplitude nonlinear internal surge in Lake Biwa. Japn. J. Limnol. **56:** 279-289.

KAWABATA, K., AND J. URABE. 1996. Population dynamics of planktonic crustacea studied during BITEX'93. Japn. J. Limnol. **57:** 545-552.

KIMURA, Y. 2001. Lake Biwa: the origin of the name. Omi Bunko (in Japanese).

KIRA, T. [ed.]. 1984. Data book of world lakes. Shiga Conference '84.

KUMAGAI, M. 2001. Seeking wisdom in limnology. Limnology. **1:** 1-2.

________, K. ISHIKAWA, AND C. JIAO. 2001a. Gyre and its related ecosystem I Lake Biwa. *In*: Kada, Y., M. Kumagai, S. Terakawa, M. Terada, T. Nakajima, and T. Nunotani [eds.], Do you know about this lake? – 50 chapters talking about Lake Biwa. Sunrise Publishing (in Japanese).

________, AND H. FUSHIMI. 1995. Inflows due to snowmelt, p.129-139. *In*: Okuda, S., J. Imberger, and M. Kumagai [eds.], Physical processes in a large lake: Lake Biwa, Japan. Coastal and Estuarine Studies. AGU Washington, D.C.

________, AND S. NAKANO. [ed.]. 1996. Baseline Data Overviews of BITEX'93. Lake Biwa Stu. Monogr. LBRI.

________, ________, C. JIAO, K. HAYAKAWA, S. TSUJIMURA, T. NAKAJIMA, J-J. FRENETTE, AND A. QUESADA. 2000. Effect of cyanobacterial blooms on thermal stratification. Limnology **1:** 191-195.

________, AND R. ROBARTS. 1996. Biwako transport experiment'93. Japn. J. Limnol. **57:** 367-558.

________, C. SHIMODA, R. TSUDA, AND T. KODAMA. 1996. Benthic and intermediate nepheloid layers in Lake Biwa. Japn. J. Limnol. **57:** 445-455.

________, T.URA, Y.KURODA, AND R.WALKER. 2001b. A new autonomous underwater vehicle designed for lake environment monitoring. Advanced Robotics. **16:** 17−26.

________, AND J. URABE. 2002. Mongolian steppe and lake. Geography. **47:** 50-55 (in Japanese).

________, ________, C.E. GOULDEN, N. SONINKHISHIG, D. HADBAATAR, S. TSUJIMURA, Y. HAYAMI, T. SEKINO, AND M. MARUO. 2003. Recent rise in water level at Lake Hövsgöl in Mongolia. *In*: Goulden, C.E., T. Sitnikova, J. Gelhaus, and B. Boldgiv [eds.], The Geology, Biodiversity and Ecology of Lake Hovsgol (Mongolia). Backhuys Belgium.

LIETH, H. 1975. Modeling the primary productivity of the world, p.237-263. *In* Lieth, H. and R.H.Whittaker [eds.], Primary productivity of the biosphere. Springer-Verlag.

MACINTYRE, S. 1996. Turbulent eddies and their implication for phytoplankton within the euphotic zone of Lake Biwa, Japan. Japn. J. Limnol. **57:** 395-410.

MASUNAGA, S., Y. YAO, I. OGURA, S. NAKAI, M. YAMAMURO, AND J. NAKANISHI. 2001. Identifying sources and mass balance of dioxin pollution in Lake Shinji Basin, Japan. Environ. Sci. Technol. **35:** 1967-1973.

MELACK, J. M., M. GASTIL, Y. AZUMA, A. HARASHIMA, AND R. TSUDA. 1996. Remote sensing of chlorophyll, suspended solids and transparency in Lake Biwa, Japan. Japn. J. Limnol. **57:** 367-375.

MORI, S., AND T. MIURA. 1990. List of plant and animal species living in Lake Biwa. Mem. Fac. Sci. Kyoto Univ., Ser. Biol. **14:** 13-32.

MURPHY, T., A. LAWSON, M. KUMAGAI, AND C. NALEWAJIKO. 2001. Release of phosphorus from sediments in Lake Biwa. Limnology **2:** 119-128.

________, M. KUMAGAI, AND K. IRVINE. 2002. The seasonal change in phosphorus dissolution in Lake Biwa sediments. Verh. Internat. Verein. Limnol. **28:** 1-3.

________, AND ________. 2002. Variation in potential for phosphorus release in Lake Biwa sediments. Yunnan Geographic Environ. Res. **14:** 41-50.

NAGATA, T., T. OGAWA, J-J. FRENETTE, L. LEGENDRE, AND W. VINCENT. 1996. Uncoupled responses of bacterial and algal production to storm-induced mixing in Lake Biwa. Japn. J. Limnol. **57:** 533-543.

NAKA, K. 1973. Secular variation of oxygen change in the deep water of Lake Biwa. Japn. J. Limnol. **34:** 41-43.

NAKANISHI, N., AND M. NAGOSHI. 1984. Yearly fluctuation of food habits of the Isaza, *Chaenogobius isaza* TANAKA , in Lake Biwa. Japn. J. Limnol. **45:** 279-288 (in

Japanese).

NAKANO, S., K. HAYAKAWA, J-J. FRENETTE, T. NAKAJIMA, C. JIAO, S. TSUJIMURA, AND M. KUMAGAI. 2001. Cyanobacterial blooms in a shallow lake: a large-scale enclosure assay to test the importance of diurnal stratification. Arch. Hydobiol. **150:** 491-509.

________, Y. SEIKE, T. SEKINO, M. OKUMURA, K. KAWABATA, M. NAKANISHI, O. MITAMURA, M. KUMAGAI, AND H. HASHITANI. 1996. A rapid growth of *Aulacoseira granulata* (Bacillariophyceae) during the typhoon season in the South Basin of Lake Biwa. Japn. J. Limnol. **57:** 493-500.

NALEWAJKO, C., AND T. MURPHY. 2001. Effects of temperature, and availability of nitrogen and phosphorus on the abundance of *Anabaena* and *Microcystis* in Lake Biwa, Japan: an experimental approach. Limnology **2:** 45-48.

NISHINO, M., M. FUKUI, AND T. NAKAJIMA. 1998. Dense mat of *Thioploca*, gliding filamentous sulfur-oxydizing bacteria in Lake Biwa, Central Japan. Water Res. **32:** 953-957.

NIPPON ASTRONOMICAL OBSERVATORY. 2002. Chronological Scientific Tables. Maruzen (in Japanese).

NIGLAS. 1990. Fuxian Lake. Ocean Press Beijing.

OGAWA, N. O., T. KOITABASHI, H. ODA, T. NAKAMURA, N. OHKOUCHI, AND E. WADA. 2001. Fluctuations of nitrogen isotope ratio of gobiid fish (Isaza) specimens and sediments in Lake Biwa, Japan, during the 20th century. Limnol. Oceanogr. **46:** 1228-1236.

OGAWA, T. AND T. NAGATA. 1996. Effects of storm disturbance on bacterial utilization of dissolved amino acids in the shallow, eutrophic South Basin of Lake Biwa. Japn. J. Limnol. **57:** 523-531.

PU, P., M. KUMAGAI, AND R. ROBARTS. [eds.]. 1998, ANSWER'97 Proceedings. J. Lake Sciences. **10:** 1-597.

RMEL. 2001, An international workshop on the restoration and management of eutrophicated lakes. Kunming, 1-227.

ROBARTS, R., M. WAISER, O. HADAS, T. ZOHAR, AND S. MACINTYRE. 1998. Relaxation of phosphorus limitation due to typhoon-induced mixing in two morphologically distinct basins of Lake Biwa, Japan. Limnol. Oceanogr. **43:** 1023-1036.

SAGGIO, A. AND J. IMBERGER. 1998. Internal wave weather in a stratified lake. Limnol. Oceanogr. **43:** 1780-1795.

SAKAI, S., S. DEGUCHI, S. URANO, H. TAKATSUKI, AND K. MEGUMI. 1999. Time trend of PCDDs/DFs in Lake Biwa and Osaka Bay. J. Environ. Chem. **9:** 379-390.

SAKAMOTO, M., AND T. INOUE. 1996. Typhoon-induced temporal change in plankton phosphorus in Lake Biwa. Japn. J. Limnol. **57:** 511-522.

SEIKE, Y., S. NAKANO, M. OKUMURA, A. HIRAYAMA, O. MITAMURA, K. FUJINAGA, M.

NAKANISHI, H. HASHITANI, AND M. KUMAGAI. 1996. Temporal variations in the nutritional state of phytoplankton communities in Lake Biwa due to typhoon. Japn. J. Limnol. **57:** 485-492.

TANAKA, Y., T. KIMOTO, AND R. TSUDA. 1996. Turbid water penetration from Yasu River into Lake Biwa at the seasonal thermocline. Japn. J. Limnol. **57:** 457-465.

________, AND R. TSUDA. 1996. Daily fluctuations in thermal stratification, chlorophyll *a*, turbidity and dissolved oxygen concentration in Lake Biwa. Japn J. Limnol. **57:** 377-393.

TODA, T. 1996. Satellite thermal remote sensing in the BITEX'93 area. Japn. J. Limnol. **57:** 553-558.

TREVORROW, M. V. 1996. Multi-frequency Acoustic Measurements of near-surface air bubbles in Lake Biwa. Japn.n. J. Limnol. **57:** 411-423.

________, AND Y. TANAKA. 1997. Acoustic and *in situ* measurements of freshwater amphipods (*Jesogammarus anandalei*) in Lake Biwa, Japan. Limnol. Oceanogr. **42:** 121-132.

URABE, J., AND K. KAWABATA. 1996. Grazing and food selection of zooplankton community in Lake Biwa during BITEX'93. Japn. J. Limnol. **57:** 27-37.

VINCENT, W. F. 1989. Cyanobacterial growth and dominance in two eutrophic lakes: review and synthesis. Ergebnisse der Limnologie **32:** 239-254.

________, M. M. GIBBS, AND R. H. SPIGEL. 1991. Eutrophication processes regulated by a plunging river inflow. Hydrobiologia **226:** 51-63.

________, C. L. VINCENT, M. T. DOWNES, AND P. J. RICHERSON. 1985. Nitrate cycling in Lake Titicaca (Peru-Bolivia): The effects of high altitude and tropicality. Freshwater Biology **15:** 31-42.

________, M. KUMAGAI, C. BELZILE, K. ISHIKAWA, AND K. HAYAKAWA. 2001. Effects of seston on ultraviolet attenuation in Lake Biwa. Limnology. **2:** 179-184.

WALKER, R., AND M. KUMAGAI. 2000. Image analysis as a tool for quantitative phycology - A computational approach to cyanobacterial taxa identification. Limnology. **1:** 107-116.

________, K. ISHIKAWA, AND M. KUMAGAI. 2002. Fluorescence-assisted image analysis of freshwater microalgae. Microbiological Methods. **51:** 149-162.

YAMADA, Y., T. UEDA, AND E. WADA. 1996. Distribution of carbon and nitrogen isotope ratios in the Yodo River watershed. Japn. J. Limnol. **57:** 467-477.

Chapter 2

Perspectives on Environmental Monitoring

2-1. Monitoring and Assessing Global Water Quality – the GEMS/Water Experience

Richard D. Robarts[1], Andrew S. Fraser[2], Kelly M. Hodgson[2], Guy M. Paquette[2]

[1] *UNEP GEMS/Water Programme Office, Environment Canada, 11 Innovation Blvd., Saskatoon, SK, CANADA S7N 3H5*

[2] *UNEP GEMS/Water Programme Office, Environment Canada, 867 Lakeshore Road, P.O. Box 5050, Burlington, Ontario, CANADA L7R 4A6*

Abstract

Evaluation and assessment of fresh and inland water quality at the regional and global scales is not a simple task. UNEPs GEMS/Water has operated a comprehensive freshwater quality monitoring and assessment programme for over 20 years and is the only such global programme. GEMS/Water operates by inviting national governments to provide water quality data from their water quality monitoring programmes. The data is then compiled into a global database, GLOWDAT, which is a value-added process. GEMS/Water, United Nations agencies and other international organizations use the data to undertake global and regional scale water quality assessments. More than 100 countries participate in the programme that has a database of >1.6 million data entries. Participating countries control, for example, the type of data collected, the location of sampling sites, the frequency of monitoring, the analytical and field methods used and the frequency at which data is transferred to GEMS/Water. In order to make effective water quality assessments, identify emerging water quality issues and environmental 'hotspots', the data available must be of good quality, comparable between countries for

Reprinted with permission from Ecohydrology & Hydrobiology 2(2002)

a specific parameter, be geographically representative for a given region and be up-to-date. The only way for GEMS/Water to ensure that all these characteristics are satisfied in GLOWDAT would be for GEMS/Water to operate its own global water quality-monitoring programme. This is economically unfeasible. However, GEMS/Water has an operational manual, a modular training course and operates a QA/QC programme to help countries with data quality. Some countries have modernized their water quality programme, a complex and comprehensive activity that includes legal and institutional considerations, technical issues, and a strategic program of capacity building. Implementation of such comprehensive programmes in more countries will lead to better quality data for GEMS/Water.

Introduction

The world's population will continue to increase even though no more fresh water will exist on earth than there was 2,000 years ago when the population was <3% of its current size (Hinrichsen et al. 1998). In many regions of the world people are removing water from rivers, lakes and aquifers faster than these systems can be recharged. It has been estimated that population growth alone will mean that the number of water-stressed, or water-scarce, countries will increase from 31 to 48 within the next 30 years (Hinrichsen et al. 1998). In addition to population growth, the demand for fresh water has been rising in response to industrial development, increasing reliance on irrigated agriculture, massive urbanization, and rising living standards (Shiklomanov 2000).

Global freshwater resources are shrinking not only in quantitative terms, but also in qualitative terms because many freshwater systems have become increasingly polluted with a wide variety of human, agricultural and industrial wastes (Shiklomanov 2000). Many developing countries are faced with difficult choices as they find themselves caught between finite and increasingly polluted water supplies on the one hand, and rapidly rising demand from population growth and development on the other (Somlyódy et al. 2001). Water shortages and pollution are causing widespread public health problems, limiting economic and agricultural development, and harming a wide range of ecosystems. Such shortfalls and contamination will put global food supplies in jeopardy and lead to economic stagnation in many areas of the world. The result could be a series of local and regional water crises with global implications (CSD 1997).

Reliable, consistent and appropriate information is the key to understanding and improving the world's supply and quality of freshwater. There is a general consensus that our knowledge of the state of the world's freshwaters needs to improve

(Shiklomanov 2000; Somlyódy et al. 2001). Freshwater and marine waters are intimately linked in the hydrological cycle so that an improvement in our knowledge of the quality of inland waters will also lead to benefits for the marine environment.

In 1998 the UN Commission for Sustainable Development called on the United Nations to undertake periodic assessments of the sustainable development, management, protection and use of freshwater resources. It requested that progress towards the universal goals adopted in Agenda 21 at the UN Earth Summit, held in Rio de Janeiro in 1992, be monitored in order to obtain a global picture of the state of freshwater resources and potential problems. The 20th Governing Council of UNEP called upon the agency to increase its knowledge of the state of the world's freshwaters. In 2000, the UN system-wide Subcommittee on Water Resources implemented the development of the World Water Assessment Programme to build upon the UN's Comprehensive Assessment of the Freshwater Resources of the World published in 1997 (CSD 1997). The 1997 report was hindered by two data related factors that undermined the conclusions:

1. The absence of reliable, comprehensive data from many countries.
2. Difficulties in assessing and comparing information from different countries and organizations.

More recently an analysis of global ecosystems concluded that data on water quality at the global level is very scarce (Revenga et al. 2000). However, there have been very few sustained programmes to monitor water quality worldwide. Currently, the only global water quality monitoring and assessment programme is UNEPs GEMS/Water.

GEMS/Water Programme

The Global Environment Monitoring System (GEMS) was inaugurated in 1972 as a result of the United Nations Stockholm Conference on the Environment. Participating governments requested that a global monitoring programme be set up to determine the status and trends of key environmental issues. The GEMS/Water Programme commenced in 1977 led by the United Nations Environment Programme (UNEP) and the World Health Organization (WHO), and assisted by the United Nations Educational, Scientific and Cultural Organization (UNESCO) and the World Meteorological Organization (WMO). The programme had two major objectives:

- The improvement of water quality monitoring and assessment capabilities in participating countries
- To determine the status and trends of regional and global water quality through the development of a global network of selected monitoring stations for lakes, reservoirs, rivers and ground waters and the compilation of a global database. By compiling a global database from multiple countries, added value was made to country-level data as it could be used to undertake global and regional scale water quality assessments.

During Phase 1 of the Programme, from 1976-1990, the global network was set up based on a data centre located at the Canada Centre for Inland Waters, Burlington, Ontario, Canada (see www.cciw.ca/gems). GEMS/Water did not actively undertake sampling programmes in countries but relied upon co-operative agreements with participating nations to provide data from their on-going water quality monitoring programmes. This continues to be the basic operating procedure and will continue to be so into the future (see below, Fig. 5). Freshwater quality assessment activities commenced at the regional and global scales using these data.

Phase 2, from 1990 to 2002, of GEMS/Water refocused the data network in order to address new global and national priorities such as land-based sources of pollution, toxic chemicals and the need for improved communication and decision skills for water quality management. New programme objectives were added to reflect international concerns on analytical methods quality control and quality assurance as well as placing a priority on improving the global representativeness of the Programme. Phase 3, for 2003 to 2013, is presently being formulated and a key new element will be improved access to the data through a revamped website.

GEMS/Water maintains direct interaction with key agencies and individuals in each participating country in a network of countries that contribute data from national, and in some cases, state monitoring programmes. The number of participating countries within GEMS/Water has risen to 103 as of 2001 (Fig. 1). Countries participate in GEMS/Water with different levels of activity and their reporting frequency is also variable. For instance, one new country joined the GEMS/Water Programme in 1996 – 97, but the increase of 21 stations during the same time period was not all associated with this country. Often countries add stations to those selected for GEMS/Water and provide data several years after it has been collected. With the growing recognition that freshwater resources are approaching a crisis at the local, regional and global scales a significant increase in activity has occurred. In addition to new countries joining the

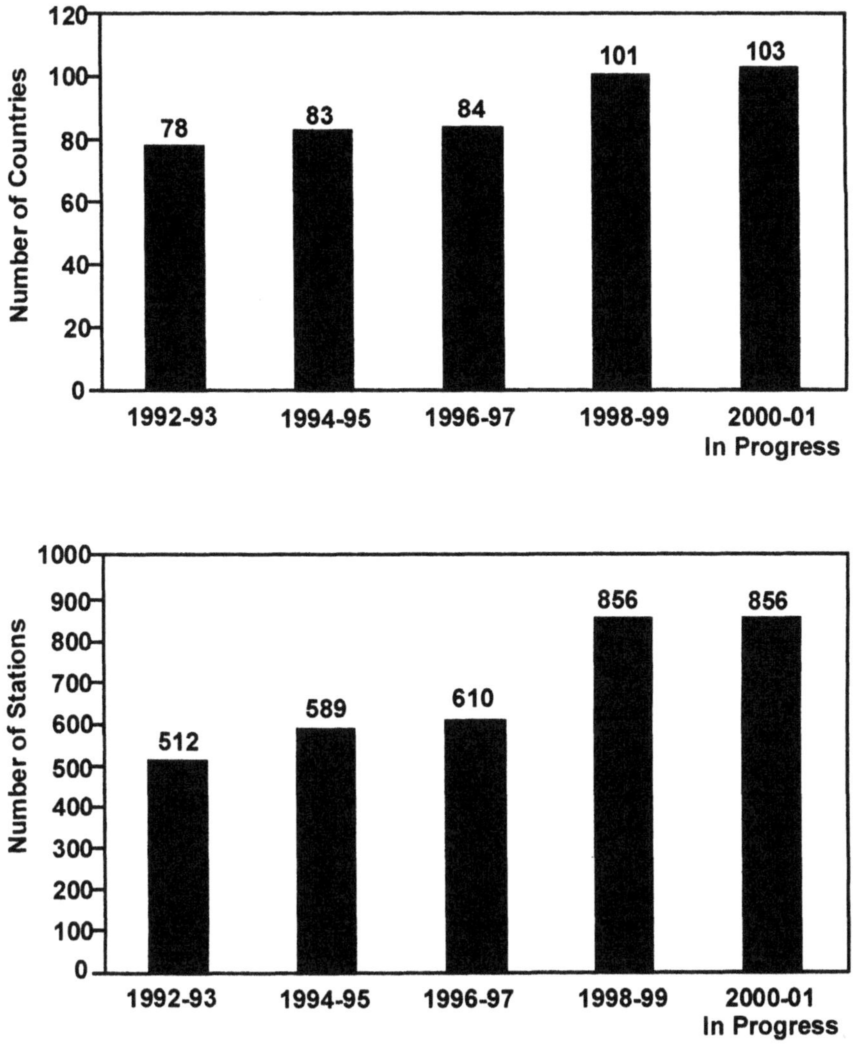

Fig. 1. The upper panel shows the change in the number of countries participating in GEMS/Water while the lower panel shows the increase in the number of monitoring stations.

GEMS/Water Programme others that have not participated for a few years are re-activating their membership and expanding their participation.

Global distribution of participating countries is widespread with the exception of one major area (Fig. 2). The continent of Africa is under represented and hence our understanding of freshwater quality conditions there is poor. UNEP has recognized the need for more information and increased technical skills in many African countries.

Fig. 2. The countries (shaded) participating in the GEMS/Water Programme as of 2001.

This has led to the development of a broad-based African initiative, which includes such programmes as Environment and Natural Resource Information Networking (ENRIN). This initiative will also include emphasizing the development of increased participation in GEMS/Water throughout Africa to help in providing the much needed data and information to determine the status and trends of freshwater resources on the continent.

Monitoring programmes in participating countries contribute data to GEMS/Water for >800 stations worldwide (Fig. 1). The number of stations for which GEMS/Water has data continues to increase and new data received are processed into the database regularly. In this way, GEMS/Water maintains a living database that is continuously updated. While the station compliment addresses the three main target sectors for freshwater resources of rivers, lakes (including reservoirs) and groundwater, unfortunately GEMS/Water does not have data for wetlands, which play important roles in the ecology and hydrological cycles of many countries. The reason for this omission is largely because wetlands are not monitored for water quality in most countries. Station data stored in the GLOWDAT database is dominated by river data followed by lakes and then groundwater. Stations are characterized into four classes:

- Baseline Stations: Located in areas where there is little or no effect from point sources of pollutants and removed from obvious anthropogenic influences.
- Impact Stations: Located at sites in which there is at least one major use of the water such as drinking water supply, irrigation, and aquatic life.

Table 1. Number of data entries by parameter group in the GEMS/Water database, GLOWDAT, for WHO designated regions: Africa, the Americas, Eastern Mediterranean, Europe, South East Asia and the Western Pacific.

Region	Physical/ Chemical	Major Ions	Metals	Nutrients	Organic Contaminants	Micro-Biology	Date Range
AFRA	2021	3916	967	1914	4	339	1978 – 1998
AMRA	33168	35226	31249	27157	3545	9384	1978 – 1999
EMRA	12257	14897	10054	8413	366	2963	1978 – 2000
EURA	112814	119061	126778	98685	11944	18968	1978 – 2000
SEAA	73186	96100	20088	53626	267	13859	1978 – 1999
WPRA	57314	39880	46627	68879	6537	9553	1978 – 2000
Total	290760	309080	235763	258674	22663	55066	1978 – 2000

Note: Table represents a sub-set of GLOWDAT encompassing major parameters only

- Trend Stations: Located primarily on large rivers that are representative of large basins in which human activity is high. Data from these stations are used mostly in regional and global assessments.
- Flux Stations: Located at the mouths of major rivers upstream from estuarine effects. Data are used to compute estimates of the load of water borne constituents from the terrestrial sector to the marine coastal sector.

As of the year 2001, the GEMS/Water database is composed of approximately 1.6 million data points (Table 1) covering over 100 water quality parameters. Classification of the data breaks into broad classes including: physical/chemical parameters, major ions, nutrients, metals, microbiological parameters, and organics. Metadata is also maintained for GEMS/Water to assist in making appropriate use of the information. Geographic distribution of the data contained in the GEMS/Water database is widespread with a higher concentration of stations in European countries, India, and Japan (Fig. 3).

Water Quality and Hydrology

In many instances, flow is the dominant component governing the health and integrity of an ecosystem and the extent and duration of pollution events in rivers. Accurate and continuing estimates of flow and hence, discharge, are essential to our capability to monitor and anticipate the impacts of the downstream transport of pollutants. In many

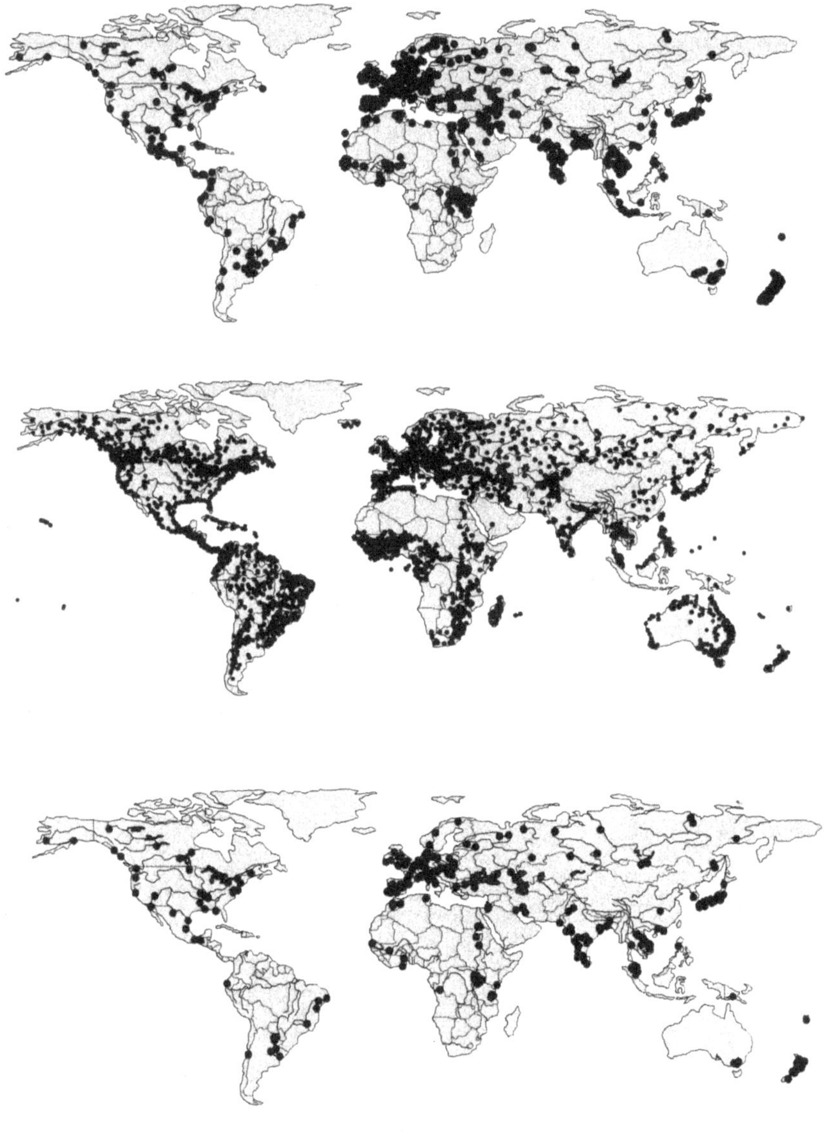

Fig. 3. The upper panel shows the distribution of GEMS/Water monitoring stations, the middle panel presents the distribution of GRDC stations and the lower panel indicates the location of joint GEMS/Water (water quality) and GRDC (hydrology) stations used for flux calculations.

cases such transport is transboundary where activities of one jurisdiction impact upon the territory of another sovereign state.

The requirements for the determination and computation of pollutant fluxes has brought about a clear recognition of the need for enhanced cooperation and interaction between the quantity and quality aspects of water resources management. The Global Runoff Data Centre (GRDC) in Koblenz, Germany, operates a global database of freshwater flows under the auspices of the World Meteorological Organization (WMO). Data are derived from 147 participating countries that operate flow-monitoring stations as an integral part of their water resources management strategies. Currently there are over 3,600 stations contributing data to the database (Fig. 3).

To link and coordinate their global water quality and quantity databases, GEMS/Water and the GRDC signed a memorandum of agreement in 1998 to begin the processes of linkage. To date the station inventories of the two programmes have been matched and a joint catalogue providing details of station characteristics is being prepared. To match stations, the geographic location was the only criteria used. Stations were usually matched with identical latitude and longitude coordinates. On occasions where the coordinates were not exactly matched a separation distance restriction of 10 km was imposed. The first round of matching criteria produced 206 joint stations (Fig. 3). The necessity of bringing the two programme elements together has required revisions to the database structures to enable ease of data storage and retrieval.

Long-term monthly averages of the relationship between flow and suspended sediment concentrations in selected rivers at joint GRDC/GEMS stations indicate the potential benefit of enhanced programme coordination. The Mekong River, which flows through China, Laos, Thailand, Cambodia, and Vietnam, exhibits highly significant wet/dry seasonality with a tight correlation to suspended sediment transport. The stretch of the river at Pakse, Laos shows the maximum suspended sediment concentration of 1526 mg L^{-1} (mean = 246 mg L^{-1}) corresponding with the maximum average monthly flow of 17,878 $m^3 s^{-1}$ in August (Fig. 4).

One of the most significant sources of suspended sediment load to the marine environment on a global scale is the Huang He (Yellow) River, China. Sediment concentrations averaging 13,000 mg L^{-1} have been observed along with high discharge rates exceeding 2,400 m^3 s^{-1} (Fig. 4). Suspended sediment concentrations are significantly correlated with discharge and are elevated in correspondence with the onset of the late summer rainy season. The region is susceptible to flooding and therefore load calculations may be significantly underestimated.

Future directions of the water quality and quantity monitoring activities of the GEMS/Water programme will emphasize the joint requirements for data acquisition

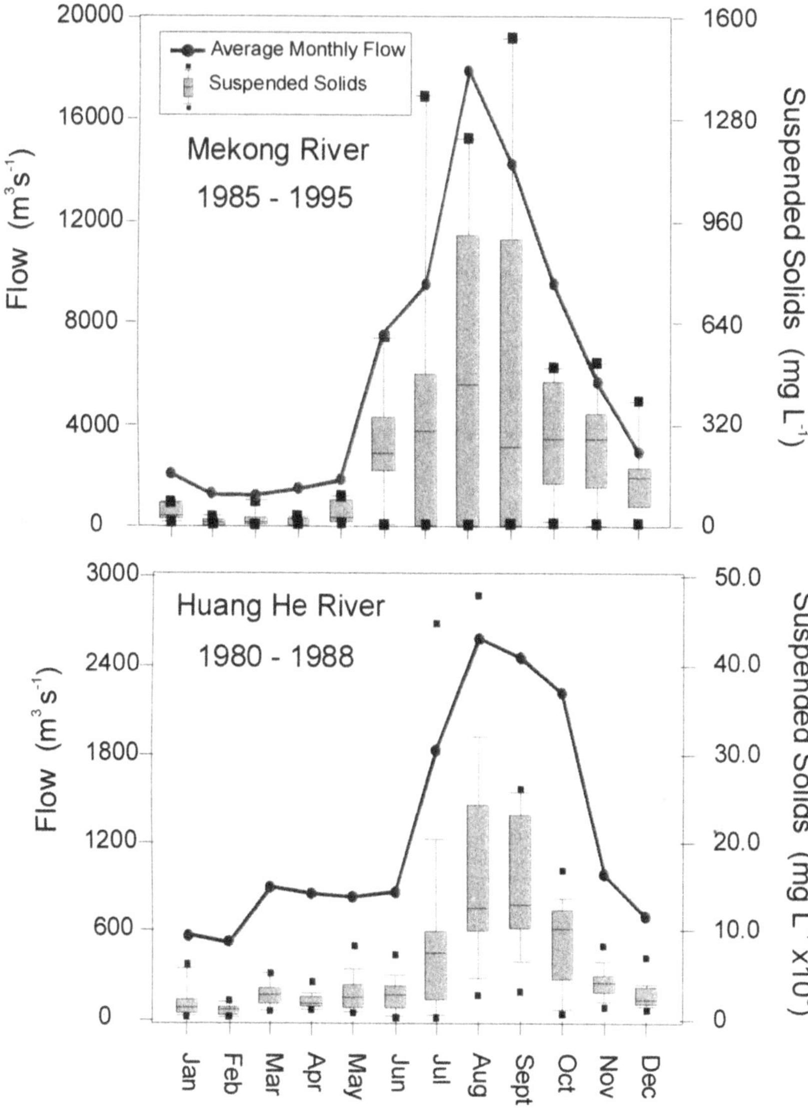

Fig. 4. Mean seasonal river flow and suspended solids concentration in the Mekong River, Laos (upper panel) and Huang He River, China. Data are plotted as Box-whisker plots showing the mean (horizontal lines insides boxes), 25th and 75th (horizontal lines on top and bottom of boxes) and 10th and 90th percentiles (horizontal lines at the ends of the vertical lines). Maxima of the frequency distribution are given as black squares.

needed for the GRDC and GEMS/Water to contribute to water quality assessments carried out by United Nations agencies and programmes (e.g., GIWA, Global

International Waters Assessment; WWAP, World Water Assessment Programme; Millennium Ecosystem Assessment; POPs Convention) and international organizations (e.g., World Bank). A major thrust will be made to provide detailed estimates of material fluxes from primary continental watersheds to the marine environment. As the meshing of the water quality and quantity monitoring activities accelerates, the overall importance of the databases, as sources for regional and global assessments, will increase.

Operational Problems for a Global Water Quality Monitoring System and Database

A variety of problems are inherent in operating a global water quality monitoring and assessment programme and generating a database of reliable information. These problems can be classified into three categories:

1. Variability of data between institutions within a country and between countries.
2. Geographical distribution of monitoring stations.
3. Time delays between the collection of data and its transfer to the global database.

Data variability

A number of factors influence the variability of data in the GEMS/Water database. These include such things as the technical capabilities of field and laboratory personnel in a country, quality of analytical equipment, methodology, and the presence of rigorous QA/QC (quality assurance and control) and laboratory accreditation programmes. Although GEMS/Water has a manual of analytical methods it recommends for use, many countries choose to employ their own methods with the result that data is compiled in GLOWDAT from a variety of methods (Table 2). In large countries, such as Russia with very extensive monitoring programmes, laboratories in different regions also may use different methods for the same parameter (Zhulidov et al. 2000b). Zhulidov et al. have noted that there has been deterioration in the quality of collected water quality data over the past decade partly because of inadequately trained personnel, poor quality of analytical equipment and reagents, and imperfect QA/QC procedures in many government laboratories. The chronic under funding of the Russian water quality monitoring system has largely been responsible for these problems and, more recently,

Table 2. Examples of the numbers of methods used per parameter and the range of method detection limits (MDL) of these methods in the GEMS/Water database.

Parameter name	Number of methods	Units	MDL range
Total boron	5	mg L^{-1}	0.002 to 40.
Dissolved boron	8	mg L^{-1}	0.02 to 0.06
Nitrate + Nitrite	6	mg L^{-1}	0.005 to 0.25
Ammonia	8	mg L^{-1}	0.001 to 0.5
Dissolved fluoride	6	mg L^{-1}	0.01 to 0.1
Orthophosphate	6	mg L^{-1}	0.0002 to 0.005
Total Phosphate	4	mg L^{-1}	0.002 to 0.005
Dissolved phosphate	2	mg L^{-1}	0.002 to 0.1
Inorganic phosphate	2	mg L^{-1}	0.002 to 0.005
Particulate phosphate	6	mg L^{-1}	0.0004 to 20
Dissolved sulphate	6	mg L^{-1}	0.01 to 5.
Dissolved chloride	7	mg L^{-1}	0.01 to 5

the loss of qualified personnel (Zhulidov et al. 2000b). Somlyódy et al. (2001) have noted that although knowledge exists, financial constraints will remain a significant problem in coming decades in most formerly socialist countries.

The impacts of these problems on Russian data quality have recently been assessed in several studies. Holmes et al. (2001) examined the long-term data sets for ammonia from the Ob' and Yenisey rivers collected by the federal Russian agency responsible for water quality monitoring, Roshydromet. The long-term average concentrations were 710 and 360 μg N L^{-1}, respectively, but they measured concentrations in both these rivers of only 10 to 15 μg N L^{-1}. While they concluded that the long-term ammonia data for these rivers were grossly in error, their preliminary conclusions about nitrate and phosphate concentrations were that these data were not as severely incorrect (Holmes et al. 2001). Since Roshydromet provides data to GEMS/Water, these errors, unfortunately, are equally applicable to the Russian data in GLOWDAT. Large discrepancies between Russian government and independent datasets have also been found for the persistent organochlorine pesticides, hexachlorocyclohexane and DDT and its derivatives, in some Russian rivers (Zhulidov et al. 2000a).

According to Ongley (2001) estimates of future levels of water pollution in many parts of the world under "business as usual" scenarios will be catastrophic for public health, the environment and for many national economies with limited resources to deal with a contaminated resource. As he and others (e.g., Somlyódy et al. 2001) have noted, the reality in many developing countries is that political and institutional instability, combined with financial restraints and poor domestic scientific capacity, means that 'western' approaches to water quality management are often inappropriate and unsustainable. Ongley (2001) addresses technical, policy, institutional, and financing issues in developing countries and proposes actions that can lead to sustainability and self-reliance for national water quality monitoring programmes. This modernization of water quality programmes achieves the twin goals of greater efficiency and greater relevance in meeting data needs for contemporary water quality management purposes in countries leading to reduced costs, less equipment and infrastructure, reduced amounts of data collected and more closely matches the abilities of developing countries to their needs (Ongley 1997, 2001). Modernization of a national water quality-monitoring programme has been successfully undertaken in Mexico (Ongley and Ordoñez 1997).

Modernization of national water quality monitoring programmes will also lead to benefits for GEMS/Water by improving the quality of data sent by participating countries. In addition to an operations manual provided to participating countries, GEMS/Water collaborates with other United Nations agencies and with participating countries in the Programme, to develop the capacity of national agencies to carry out water quality monitoring and assessment activities in their countries by offering training programmes of analytical quality assurance, information systems, computer based analysis systems, and water quality management programme strategies. In addition, GEMS/Water operates a QA/QC programme for participating laboratories to help them improve the accuracy and precision of their results. In Phase 3 of GEMS/Water, the QA/QC programme will play a stronger role than has historically been the case in helping provide reliable data. Recommended quality control criteria for analytical methods, where precision, accuracy, contamination, recovery and stability are continually monitored to ensure reliability of data, will be provided to laboratories. An Analytical Methods Dictionary for all methods used in GEMS/Water has also been drafted. The Dictionary provides information on the analytical principle for each method, the equipment required, the Method Detection Limit, and literature references, which will be very useful for both participating laboratories and GEMS/Water in comparing the performance of different analytical methods.

Geographic coverage

GEMS/Water through its network of National Focal Points (NFP = person in each country nominated by their government to be the official GEMS/Water contact) is working to increase the distribution of stations worldwide to help improve the representativeness of the data and information in GLOWDAT. The Programme has moved away from the policy of the past where only data for stations designated as GEMS/Water stations is compiled. Instead, GEMS/Water accepts data from all stations within a country that are being monitored. In this way, maps and assessments of water quality at the global or regional scales will be more effective and reliable. While it is unlikely that GEMS/Water will ever have the equivalent geographic coverage of stations that the GRDC has for hydrological data (see below, Fig. 3) significant increases in geographic coverage will occur. For example, the number of stations for Canada should increase from the current 17 to about 100 and for the United States, from the present 21 to about 300.

Time delays in transferring data

Water quality data that are made available to users soon after water samples have been collected and analyzed can be used to identify emerging water issues and environmental "hotspots", a key concern for UNEP. In order to ensure minimal delay between data production in a country and its receipt by GEMS/Water, GEMS/Water places a high value on strong communication with each participating country's NFP. Conditions and infrastructure for every country vary widely and this adds considerably to the complexity of strong and effective communication. Experience has shown that the flow of data and information transferred to GEMS/Water is proportional to the level of communications with the NFP for a country. Support for activity in GEMS/Water by a country is likely to be reduced if feedback from GEMS/Water on their participation is not received in a timely manner.

Electronic based communications through the Internet and the use of email are now the main procedures for communications in GEMS/Water. Not all countries are able to support Internet activity so GEMS/Water must maintain all formats. Approximately 20% of countries currently use electronic transfer protocols. An objective of GEMS/Water Phase 3 is to increase the capability for all participating countries to exchange information via the Web.

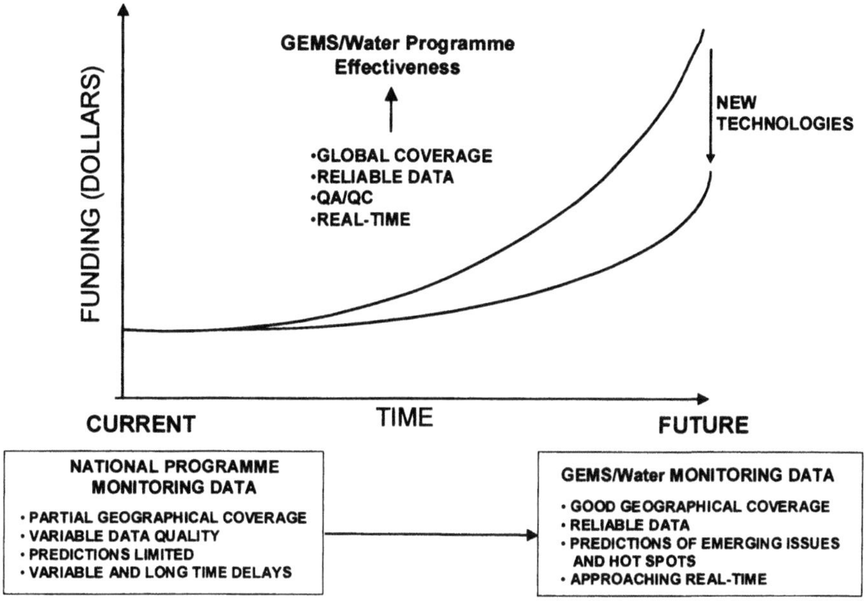

Fig. 5. Schematic diagram of the dilemma facing the GEMS/Water Programme – the need for more reliable and timely data from an effective geographical distribution of monitoring stations versus raising costs to obtain it. Further explanation is provided in the text.

Resolution of Problems and Financial Implications

The financial consequences of resolving, or at least minimizing, all the problems in compiling a global water quality database are potentially staggering (Fig. 5). While training programmes and a rigorous QA/QC activity will lead to improvements and help maintain the quality of data, GEMS/Water remains reliant on the goodwill and strong commitment of its participating countries for data. Hence, issues of data quality, geographic coverage of stations, the choice of parameters monitored and the frequency at which data is conveyed to GEMS/Water, are the prerogatives of national governments, not GEMS/Water. Although it is an unrealistic concept, at least for the foreseeable future, the best solution to these problems would be if GEMS/Water could operate its own monitoring programme throughout the world. The cost to do this would be enormous and there would be huge political and institutional problems that probably could not all be resolved. Future technological developments, where electronic sensors could be inserted into waters to monitor a broad suite of water quality and quantity parameters that would be uploaded to a satellite and then downloaded to GEMS/Water are at the moment largely in the realm of science fiction. But potentially such developments could significantly reduce the cost of running a global water quality-monitoring network. Already sensors are available to do this for very basic water quality parameters, such as dissolved oxygen, conductivity, water temperature and, to some extent, nutrients (phosphorus and nitrogen concentrations). Consequently, until this technological future arrives, GEMS/Water will continue to operate a global water quality and assessment programme relying on national governments. This means that the problems of data quality, geographical coverage and time lags will not be resolved for the foreseeable future. But with capacity building, QA/QC and programme modernization within developing countries, the impacts of these should decrease appreciably with time. In addition, GEMS/Water will strengthen its collaboration with other programmes and projects from which it might be able to draw information and data. However, it cannot be overemphasized that the success of GEMS/Water is largely dependent on the goodwill and commitment of national governments worldwide.

Conclusion

Clearly the need for reliable global information on freshwater quality has not diminished since the inception of GEMS/Water more than 20 years ago, but has grown. While GEMS/Water remains the only global water quality monitoring and assessment programme, there are critical issues that need to be addressed in terms of the effectiveness and reliability of the Programme's database. Capacity building in developing countries, an effective QA/QC programme, water quality monitoring programme modernization within countries, combined with the strong commitment of governments to support national water quality monitoring programmes and to participate in GEMS/Water, will all contribute to minimizing the impacts of these issues on data quality. The GEMS/Water database is a living database that is unique and urgently needed if we are to have the information that is essential for assessing the state, and managing the quality, of the world's inland aquatic systems.

Acknowledgements

We thank Dr. Salif Diop and P. Mmayi of the Division of Early Warning and Assessment, UNEP, Nairobi for comments on a draft of this manuscript.

References

CSD (Commission for Sustainable Development). 1997. Comprehensive assessment of the freshwater resources of the world. Stockholm Environment Institute and World Meteorological Organization, Geneva. 35 p.

HINRICHSEN, D., B. ROBEY, AND U. D. UPADHYAY. 1998. Solutions for a water-short world. Population Reports, Series M, No. **14**, John Hopkins School of Public Health, Population Information Program, Baltimore. 31 p.

HOLMES, R. M., B. J. PETERSON, A. V. ZHULIDOV, V. V. GORDEEV, P. N. MAKKAVEEV, P. A. STUNZHAS, L. S. KOSMENKO, G. H. KÖHLER, AND A. I. SHIKLOMANOV. 2001. Nutrient chemistry of the Ob' and Yenisey Rivers, Siberia: Results from June 2000 expedition and evaluation of long-term data sets. Mar. Chem. **75:**219-227.

ONGLEY, E. D. 1997. Matching water quality programs to management needs in developing countries: The challenge of program modernization. Euro. Wat. Pollut. Cont. **7:**43-48.

_______. 2001. Water quality programs in developing countries: Design, capacity building, financing and sustainability. Wat. Internat. **26:**14-23.

_______, AND E. B. ORDOÑEZ. 1997. Redesign and modernization of the Mexican water quality monitoring network. Wat. Internat. **22:**187-194.

REVENGA, C., J. BRUNNER, N. HENNINGER, K. KASSEM, AND R. PAYNE. 2000. Pilot analysis of freshwater ecosystems. Freshwater systems. World Resources Institute, Washington, D.C.

SHIKLOMANOV, I. A. 2000. Appraisal and assessment of world water resources. Wat. Internat. **25:**11-32.

SOMLYÓDY, L., D. YATES, AND O. VARIS. 2001. Challenges to freshwater management. Ecohydrol. Hydrobiol. **1:**65-95.

ZHULIDOV, A.V., J. V. HEADLEY, D. F. PAVLOV, R. D. ROBARTS, L. G. KOROTOVA, V. V. VINNIKOV, AND O. V. ZHULIDOVA. 2000a. Riverine fluxes of the persistent organochlorine pesticides hexachlorocyclohexane and DDT in the Russian Federation. Chemosphere. **41:**829-841.

_______, V. V. KHLOBYSTOV, R. D. ROBARTS, AND D. F. PAVLOV. 2000b. Critical analysis of water quality monitoring in the Russian Federation and former Soviet Union. Can. J. Fish. Aquat. Sci. **57:**1932-1939.

2-2. Bio-Optical Variability in the Littoral Zone: Local Heterogeneity and Implications for Water Quality Monitoring

Jean-Jacques Frenette[1] and Warwick F. Vincent[2]

[1]*Département de Chimie-Biologie, Université du Québec à Trois-Rivières*
C.P. 500, Trois-Rivières, QC, G9A 5H7, Canada
[2] *Département de Biologie, Université Laval, Québec, QC, G1K 7P4, Canada*

Abstract

The understanding of local, small-scale heterogeneity is critical to the successful monitoring and management of whole lake ecosystems. This variability may involve differences in physical and chemical characteristics which in turn control biological community structure, biodiversity and local water quality. The littoral zone typically shows the highest productivity per unit area because of the high availability of light for photosynthesis in the water column and the transfer of nutrients from the adjacent watershed. Although light exerts a controlling influence on aquatic ecosystems, little information is available about the bio-optical variability of inshore waters. This variability is likely to generate increased physical complexity, favouring a broad range of species. Optical factors may also operate as selective pressures for organisms living in the littoral zone due to the greater UV exposure and other changes in the underwater spectral composition. In this chapter, we examine the factors controlling light in the littoral zone as a key component and also index of habitat variability in lakes. We address the potential to use bio-optical criteria and technologies as a guide to habitat variability and for water quality monitoring. Optical measurements now offer enormous potential for the study of algal blooms, eutrophication and shifts in plankton community structure. We illustrate the application of bio-optical analysis by way of our recent results from the south and north basins of Lake Biwa, Japan, and from Lake Saint-Pierre, Canada, a shallow fluvial lake that experiences seasonal flooding.

Introduction

The littoral zone plays a major role in the functioning of lake ecosystems and is often the region where water quality issues are of greatest concern. It is the zone of highest productivity per unit area because of the high proportion of light available for photosynthesis in the water column and the transfer of carbon and nutrients from the adjacent terrestrial ecosystem (Wetzel 2001). Furthermore, the inshore region is typically the site of most recreational activity and of the greatest human impacts. It is also the site where water blooms accumulate and have the strongest influence on human perception of lakewater quality. For example, cyanobacterial blooms of buoyancy regulating species (especially the genera *Microcystis, Anabaena* and *Aphanizomenon*) are often initiated in shallow and productive waters where thermal stratification of the water column occurs diurnally or occasionally persists for longer periods (Oliver and Ganf 2000). Such habitats are typical of calm and sheltered areas of the littoral zone and can act as incubation sites for water bloom organisms which are then advected out into the pelagic zone during windy days (Ishikawa et al. 2002).

Water quality monitoring is undertaken for two purposes: scientific research and water quality management. These two types of monitoring programmes serve different objectives but share some similar needs. In both sets of objectives, for example, there is a need to select a representative sampling site for observations in the lake. However, despite the pre-eminent importance of the littoral zone, water quality monitoring strategies typically emphasize offshore sites. A central basin location is often considered representative of the greater volume of the lake, yet it is unlikely to reflect accurately (or at all) the primary sites of water quality impacts and concerns.

One of the impediments to routine monitoring of the inshore environment is its great spatial variability. This heterogeneity precludes many types of limnological methods because of time, cost and logistic constraints, and it results in the need for rapid approaches that can be applied to many sites. Bio-optical technologies are now emerging that are beginning to address these needs. Bio-optical instrumentation for underwater applications, for example in terms of spectral resolution and automated logging capabilities, hold considerable promise for water quality analysis and surveillance.

Measurements based on underwater optics provide not only a guide to local-scale variability in water quality, but also provide fundamental information about habitat characteristics of the lake given the critical role of solar radiation in photochemistry and photobiology, especially primary production of the phytoplankton and macrophytes, and also in underwater vision and behavior by aquatic animals (e.g., Seehausen et al. 1997). Underwater light can also operate in more subtle ways for example in controlling algal stoichiometry and lipid composition that in turn have implications for food quality and transfer to higher trophic levels (Wainman et al. 1999; Sterner et al. 1999; Frenette et al. 1999).

In this chapter we explore the factors controlling the inshore irradiance regime and the potential of using bio-optical measurements to characterize the littoral environment. We focus on local-scale variability and its implications for habitat diversity and monitoring and illustrate these aspects by way of two examples from our recent research: Lake Biwa, Japan, and Lake Saint-Pierre, Québec, Canada. These two lakes are both designated UNESCO world heritage ecological reserves.

Material and Methods

Study sites

Lake Biwa

Lake Biwa is an ancient lake that formed as a tectonic basin some 4 million years ago (Nakajima and Nakai 1994). It is characterized by a rich biodiversity with 491 aquatic plants (including phytoplankton) and 595 animal species (Mori and Miura 1990). It is the habitat for more than 54 endemic species (Nakajima and Nakai 1994). Lake Biwa provides drinking water to 14 million people but since 1970 has experienced a deterioration in water quality due to blooms of nuisance planktonic algae (Nakanishi and Sekino 1996). Concern has been expressed about the continuing degradation of all regions of the lake with special concern about the exploitation of the littoral zone. A great part of the shoreline of the lake has been changed and simplified artificially by the Comprehensive Development Project of Lake Biwa proclaimed in 1971 (Nakanishi and Sekino 1996). Among other impacts, this resulted in the loss by drainage and reclamation of aquatic environments across more than 20% of the whole littoral area. A substantial increase in population density in the Lake Biwa basin (from 1.14 million in 1984 to 1.27 million in 1994) has also increased the environmental pressure on these inshore waters.

Lake Biwa is characterized by two basins that differ greatly in their physical and chemical properties (Fig. 1). The North Basin is large (620 km^2), deep (104 m maximum depth; mean depth of 44 m) and monomictic. The South Basin is smaller (57 km^2), shallow (8 m in maximum depth; mean depth 2.8 m) and polymictic.

This paper concentrates on the analysis of a mid summer data set obtained on August 24, 1998. In the broader program, sampling was conducted every 2-3 weeks throughout the major productivity season of the lake. Cruises were conducted on seven dates at 11-16 stations located using a differential Global Positioning System (dGPS) with ± 2 m accuracy, distributed throughout the lake from June 1 to October 5, 1998. Three stations were located in the South Basin and 8 stations were in the North Basin. Two of the stations were in the pelagic offshore waters while all of the remaining stations were located within the littoral zone with depths ranging from 2.5 to 7.5 m.

Lake Saint-Pierre

Lake Saint-Pierre (lat: 46°12'; long: 72°50') is the largest (315 km^2) fluvial lake along the St. Lawrence River and the last major enlargement (13.1 km width at mean discharge) of the river before the estuary (see Vincent and Dodson 1999). It is typically shallow (average depth of 3.17 m at mean discharge) and covered with extensive macrophyte beds during summer. It has recently been classified (2000) by UNESCO as an "ecological reserve of the biosphere" and is thus now officially recognized as a world heritage site. This recognition arose, in part, because of its high biodiversity; the lake supports 83 fish species and 288 bird species (Langlois et al. 1992).

The lake is characterized by three water masses which are composed of inflows from various tributaries which vary in their flow rates and concentrations of dissolved and particulate organic and inorganic suspended matter. The relative contribution of these different water masses to the physico-chemical makeup of Lake Saint-Pierre fluctuates with the relative inputs of the major inflows and smaller tributaries to total discharge. Further information is given in Frenette et al. (2002).

This paper concentrates on the analysis of an early summer data set obtained on June 19, 2000, at the beginning of the macrophyte growing season. In the broader program, sampling was conducted every 2-3 weeks throughout the major productivity season of the lake. Cruises were conducted on seven dates at 27 stations distributed in four transects throughout the lake from June 19 to September 14. Of these, 19 stations (Sites 5 to 23) were located using dGPS (± 2 m accuracy), distributed along two transects located perpendicular to the main axis of the North and South shores of Lake Saint-Pierre and spanning all three water masses (Fig. 2; details in Frenette et al. 2002).

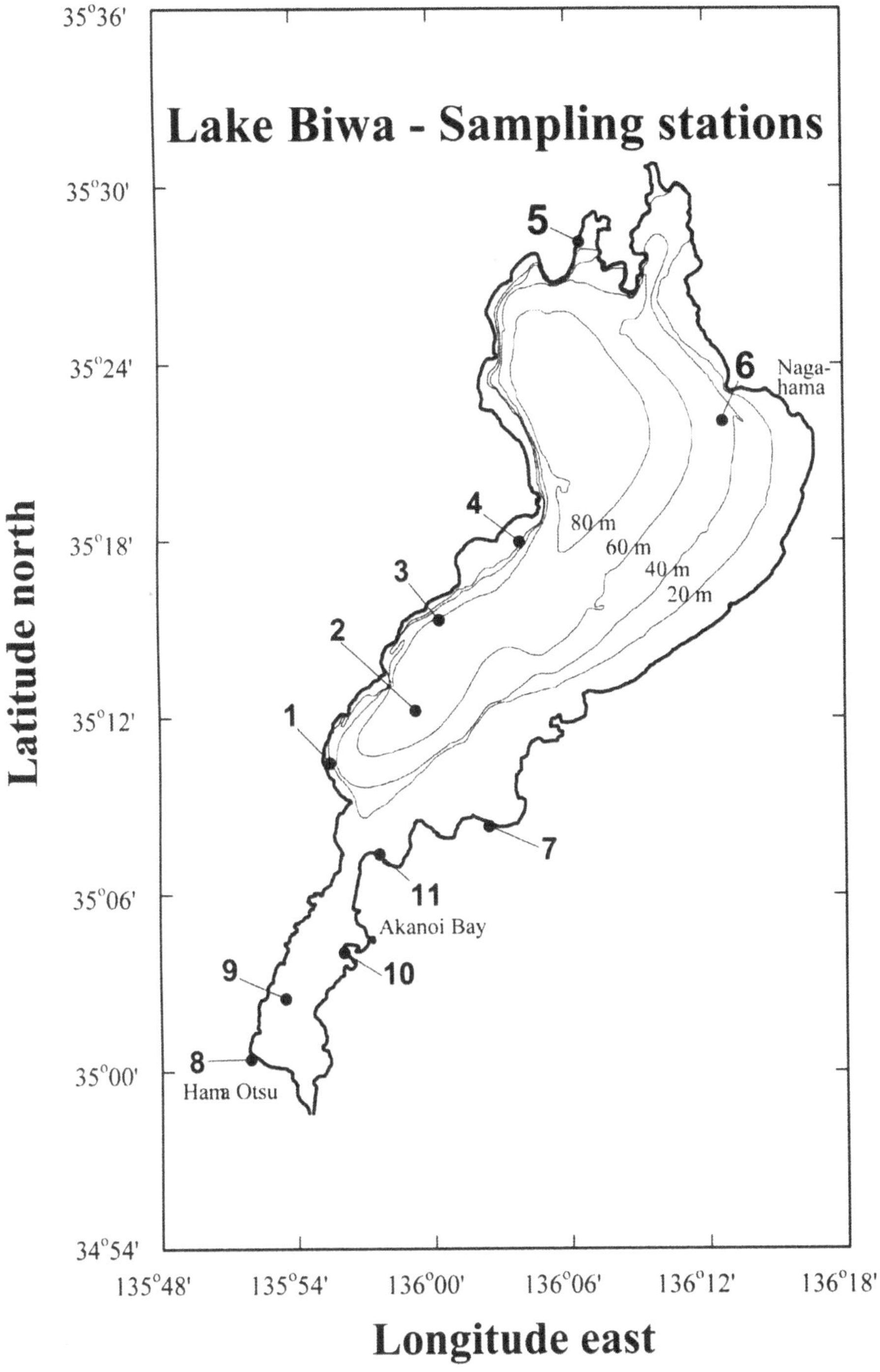

Fig. 1. Map of Lake Biwa showing the position of sampling stations on August 24, 1998.

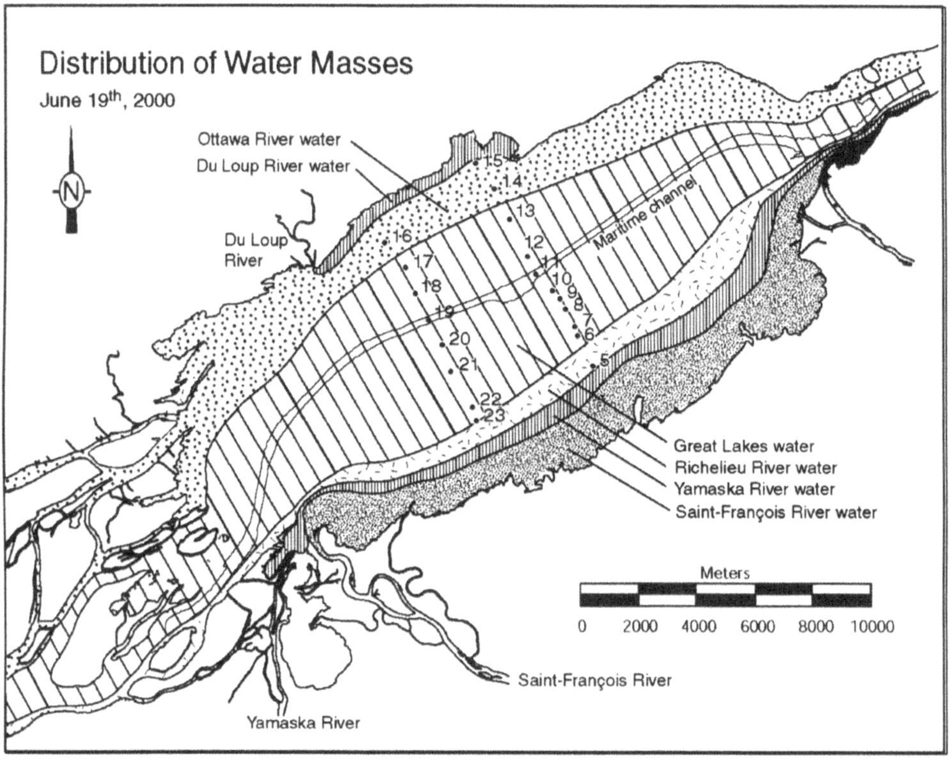

Fig. 2. Distribution of water masses in Lake Saint Pierre, Quebec. Reproduced from Frenette et al (2002) with permision from Kluwers.

Optical approaches

In our studies at Lake Biwa, transmittance, ultraviolet radiation (UV) and photosynthetically available radiation (PAR) were measured by way of a Biospherical PUV-550 (San Diego, USA) profiling radiometer fitted with a Wet Labs C-star transmissiometer (488 nm). The profiler measured cosine-corrected downwelling underwater irradiance (E_d) at 313, 340, 443 and 550 nm in addition to full cosine-corrected PAR. In Lake Saint-Pierre, a Biospherical PPR-800 spectroradiometer was used to measure the cosine-corrected downwelling underwater irradiance (E_d) at 340, 380, 395, 412, 443, 465, 490, 510, 532, 560, 565, 589, 625, 665, 670, 683, 694, 710 nm and downwelling cosine-corrected PAR (400-700 nm). The instruments were slowly lowered through from the surface to the bottom of the water column at each station and more than 100 measurements·m^{-1} were recorded on a portable computer. Data were corrected by subtracting the "dark irradiance" values (obtained when the instrument was fitted with a light-tight neoprene cap at *in situ* temperatures) from the $E_d(\lambda)$ readings. Diffuse vertical attenuation coefficients (K_d) were calculated by linear regression of the natural logarithm of E_d versus depth. The depth to which 1% of subsurface irradiance penetrated ($Z_{1\%}$) was calculated as 4.605/K_d (Kirk 1994). UVA and UVB radiation were measured at 340 nm and 313 nm respectively to calculate the diffuse attenuation coefficients K_{d340} and K_{d313}. An index of UV exposure throughout the water column was calculated as ($Z_{1\%UV}$)/ Z_{max} where $Z_{1\%UV}$ is the depth of 1% of subsurface irradiance at 313 and 340 nm and Z_{max} is the depth of the water column.

RESULTS

Lake Biwa

Waveband penetration

The most penetrating wavebands were green (550 nm), followed by blue (443 nm), followed by UVA (340 nm) then UVB (313 nm) irradiance, as is typical of Case 2 waters containing colored dissolved organic matter (CDOM) from terrestrial vegetation sources (Fig. 3). The euphotic zone (1% PAR) extended to the bottom of the water column at every station of the littoral zone due to the shallow depths and relatively high water clarity throughout the lake. In the pelagic zone, the 1% PAR depth extended to 21% and 96% of the water column in the North and South Basins respectively. A typical set of PUV 545 profiles for station 5 (the northernmost station in the North Basin) had K_d values of 1.83 m^{-1} (313 nm), 1.29 m^{-1} (340 nm), 0.53 m^{-1} (443 nm), 0.21 m^{-1} (550 nm) and 0.36 m^{-1} (PAR) (Fig. 3).

North versus South Basin of Lake Biwa

Consistent with previous studies on the pelagic zone of Lake Biwa, the 1% depth penetration for PAR (Tsuda and Nakanishi 1988) and also for UVR (Vincent et al. 2001) was much greater in the offshore waters of the North relative to the South Basin (Fig. 4). The same pattern also held for the littoral zone. The UVB (313 nm) 1% penetration ranged from 0.6 to 1.3 m in the South Basin and from 1.6 to 2.7 m in the North Basin. The equivalent values for UVA (340 nm) were 0.66 and 1.4 m in the South Basin and 1.6 and 3.5 m in the North Basin. For PAR, values varied between 2.8 and 4.3 m for the South Basin and 5.6 and 12.8 m for the North Basin.

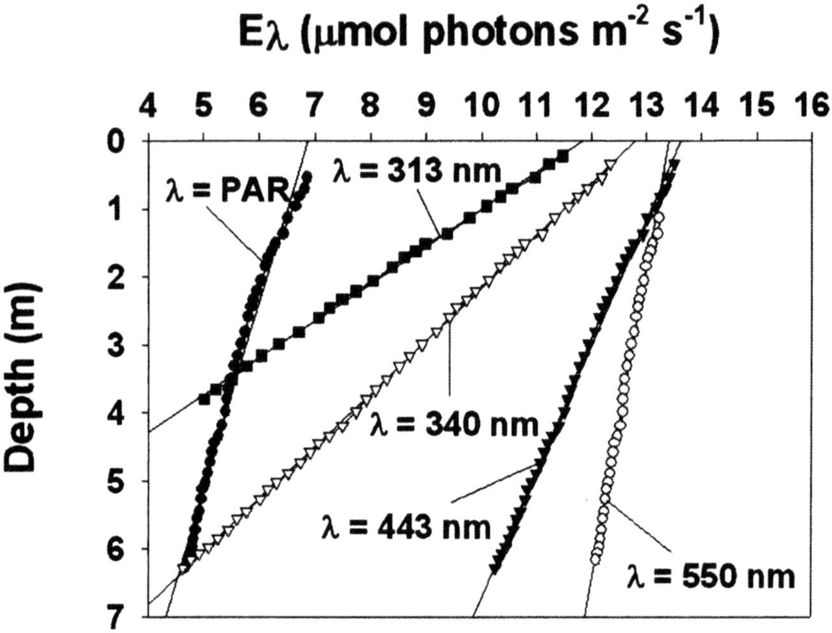

Fig. 3. Penetration of PAR, blue (443 nm), green (550 nm) and UVR at two wavelengths (313 and 340 nm). The measurements were made around midday at Station 5 (northernmost station in North Basin), August 24, 1998. The attenuation curves are from log-linear regressions.

Littoral versus pelagic

The transparency of the water column to all wavebands was always reduced in the North Basin littoral stations relative to the offshore reference site (Fig. 5). Water transparency was consistently higher in Hama Otsu (South Basin) than the mid-South Basin Site. The lowest transparencies throughout the lake were always recorded in Akanoi Bay (South Basin). Average PAR irradiance values in the water column (Kirk 1994) were consistently higher in the littoral than the pelagic zone

Littoral variability

In the littoral zone, the UVR generally penetrated a large proportion of the water column with 32 to 62% of the water column exposed to more than 1% subsurface UVB irradiance in the North Basin and 30 to 46% of the South Basin water column (Fig. 6). In the pelagic

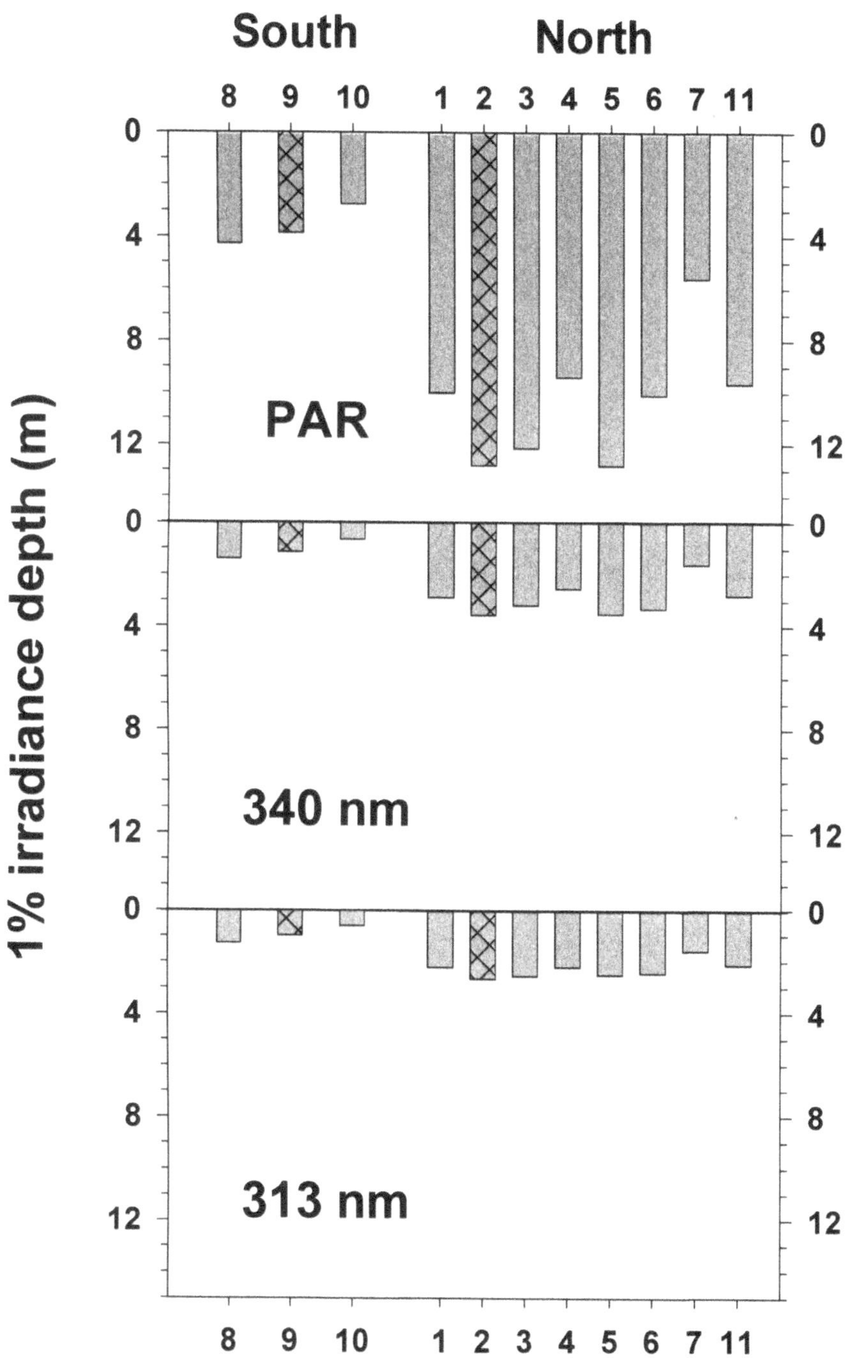

Fig. 4. 1% penetration depth for PAR (400-700 nm), UVB (313 nm) and UVA (340 nm) at each sampled site in the North and South Basins of Lake Biwa. Hatched areas refer to the pelagic zones of each basin.

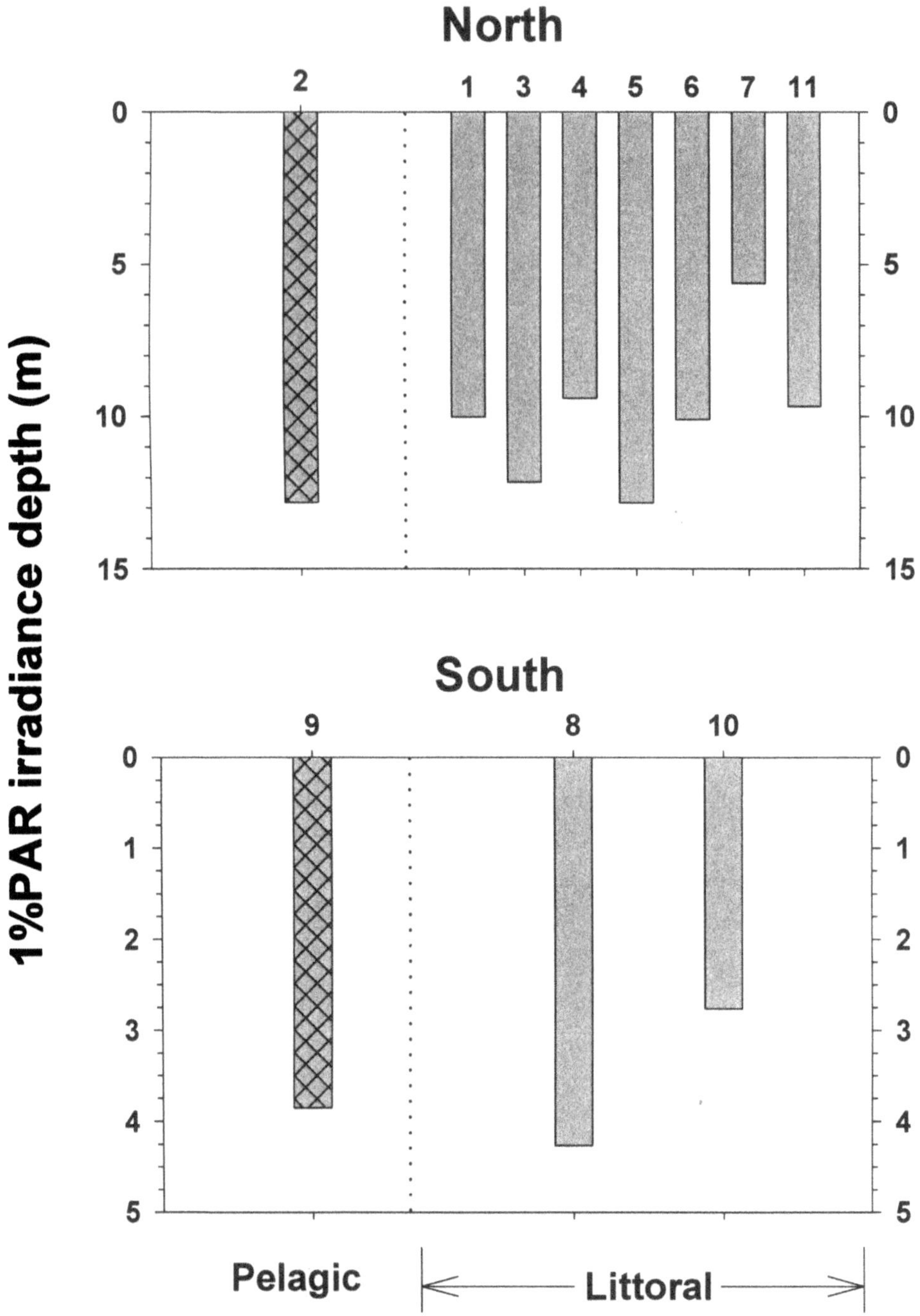

Fig. 5. 1% PAR (400-700 nm) irradiance depth for the littoral and pelagic zones in the North and South Basins of Lake Biwa. Hatched areas refer to the pelagic zones of each basin.

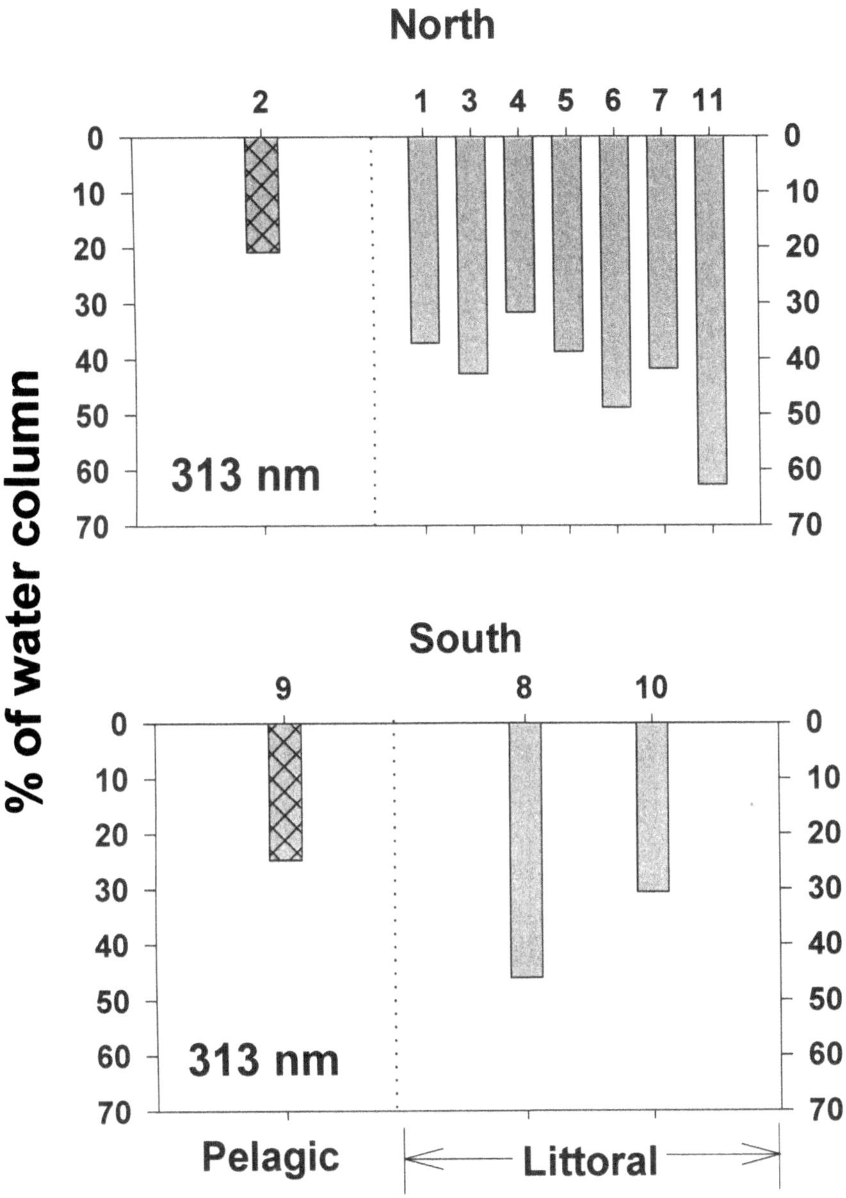

Fig. 6. 1% UVR depth penetration in the whole water column expressed as the percentage of the whole water column for the littoral and pelagic zones in the North and South Basins of Lake Biwa. Hatched areas refer to the pelagic zones of each basin.

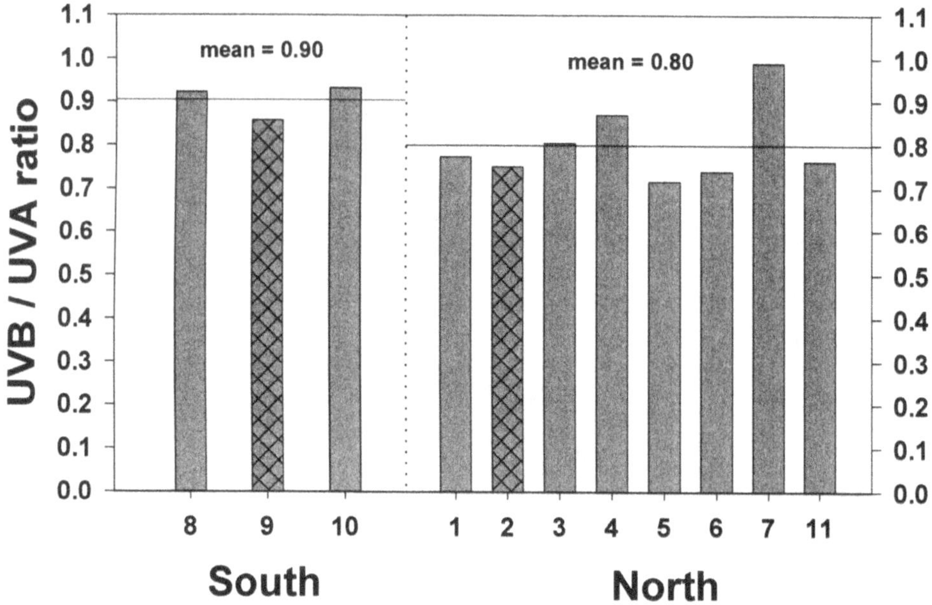

Fig. 7. UVB/UVA ratio expressed as K_d340/K_d313 where K_d340 and K_d313 are the attenuation coefficients at 340 and 313 nm of the whole water column for the littoral and pelagic zones in the North and South Basins of Lake Biwa. Hatched areas refer to the pelagic zones of each basin.

zone, UVR penetration expressed as a % of the euphotic zone reached 21% of the 60 m deep water column of the North Basin (Fig. 3). There was also considerable variation throughout the lake in the relative penetration of UVB versus UVA wavebands (Fig. 7).

There was significant variability between most (east and west) of the North Basin sites in terms of 1% penetration depth for UVR. For stations located on the west side (1-6), values varied within 20% of mean. However penetration was typically less on the east side of the North Basin (stations 7 and 11) with variations within 33% of mean. These sites are located near the mouth of tributaries (e.g., Yasu River). Values for the South Basin sites showed a much greater variability in the littoral zone with factor of two differences in the 1% irradiance depth between Hama Otsu and Akanoi Bay.

Lake Saint-Pierre

Variability within the three water masses

The three water masses varied markedly in spectral attenuation and transparency (Fig. 8).

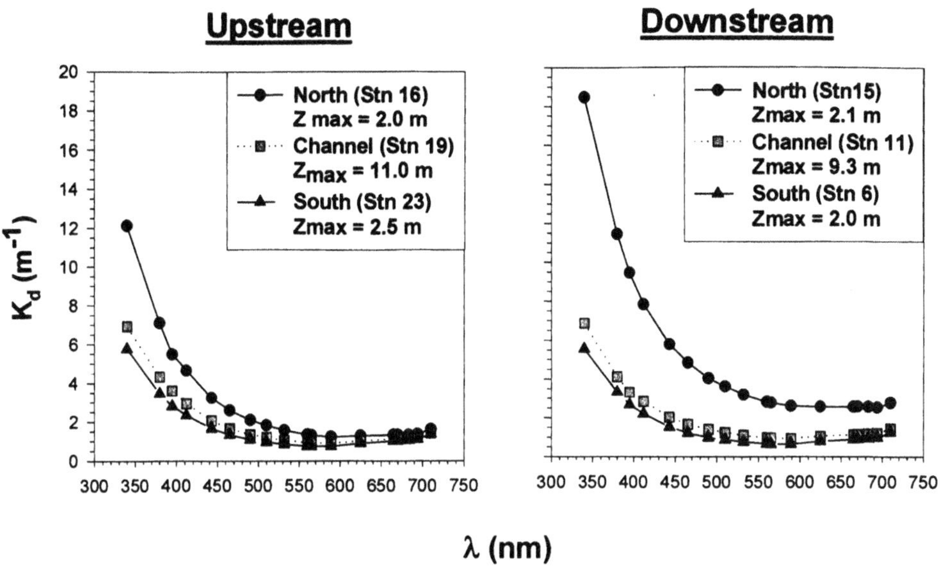

Fig. 8. Wavelength-dependent light penetration expressed as the cosine-corrected downwelling attenuation coefficients (K_d) at 340, 380, 395, 412, 443, 465, 490, 510, 532, 560, 565, 589, 625, 665, 670, 683, 694,and 710 nm at three stations representative of the north, central and south water masses in Lake Saint-Pierre for both the upstream (i.e. stations 16 to 23) and downstream (i.e. stations 5 to 15) transects, panels A and B, respectively. Reproduced from Frenette et al. (2002) with permision from Kluwers.

For example, the northern water mass (e.g., stations 16 & 15) was the most turbid with the red-green portion of the spectrum being the most penetrating. The southern water mass (e.g., stations 23 & 6) tended to be less turbid, and had higher PAR penetration with the most penetrating waveband in the green part (550-600 nm) of the spectrum. The transparency of the central water mass (e.g., stations 19 & 11) was intermediate between the north and south water masses (Fig. 8). Further, the depth to which 1% of surface irradiance penetrated ($Z_{1\%}$) differed substantially amongst the three water masses (Fig. 9). For the upstream and downstream portion of the study area, the whole water column was exposed to more than 1% of surface PAR at all stations except in the maritime channel where the euphotic depth (1% PAR) extended over 42-47% of the water column (Fig. 9). Further details are given in Frenette et al. (2002).

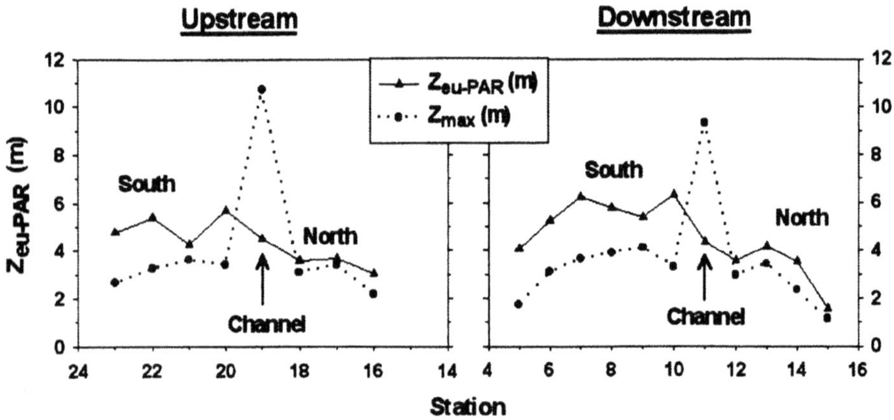

Fig. 9. Euphotic zone (Z_{eu}PAR) and maximum depth (Z_{max}) along the north-south axis at both the upstream (panels A & C, representing stations 16 to 23) and downstream (panels B & D, representing stations 5 to 15) transects. Reproduced from Frenette et al (2002) with permission from Kluwers.

Table 1. Effects of a macrophyte bed on light penetration depth (UV-A and PAR) and dissolved and particulate organic carbon concentrations in the water column of Lake Saint-Pierre, August 28, 2001. The Z values refer to the 1% surface irradiance depth for UV-A (340 nm) and PAR (400-700 nm).

Distance from the river (km)	$Z_{1\%\ UV\text{-}A}$ (m)	$Z_{1\%\ PAR}$ (m)	DOC (mg l^{-1})	POC (mg l^{-1})
4.0	0.42	2.05	4.05	0.88
4.5	0.42	1.76	4.31	1.07
5.0	0.42	1.84	4.76	0.54
5.5	0.56	1.39	4.14	0.35
6.0	0.63	0.76	3.99	0.22
6.5	0.66	1.16	4.09	0.27
7.0	0.68	1.06	3.74	0.30
7.5	0.74	0.73	3.48	0.31
8.0	0.95	0.84	3.41	0.22
8.5	1.14	0.53	3.03	0.18

Upstream-downstream variability

Over a 5 km transect, macrophytes had a profound influence on the optical characteristics of the water mass from a nearby tributary circulating through the beds. Macrophytes were responsible for a decrease in both particulate matter and CDOM with an increasing penetration depth of UVR and PAR in the water column (Table 1) (Martin and Frenette in prep).

DISCUSSION

Measurements of spatial variability of optical properties throughout lakes are relatively rare (for notable exceptions see Effler et al. 1991; Seehausen et al. 1997; Smith et al. 1999 and Belzile et al. 2002). In general, optical data sets do not allow systematic comparisons between stations in the littoral zone which are often undersampled compared to pelagic stations. Our results from Lake Biwa and Lake Saint-Pierre draw attention to the large spatial variability in underwater conditions that may be encountered in the lacustrine environment, and the utility of bio-optical approaches in defining the magnitude of site-to-site variation. This variability was greatest among littoral (inshore) sites, determined in part by variations in the inherent optical properties of inflowing tributaries, as well as the presence of macrophytes, the morphometry of the shoreline such as embayments and variations in depth, and the extent of wind-induced suspension of light-attenuating sediments (see Belzile et al. 2002). This variability was expressed in terms of PAR and UV attenuation, as well as in terms of waveband ratios. For example, the ratio of UVB radiation relative to UVA was generally greater in the South than the North Basin of Lake Biwa, and in the Southshore relative to Northshore water mass of Lake Saint-Pierre (Fig. 7). This indicates that photosynthetic organisms may have less facility to recover from UV-induced damage due to less UVA radiation that is known to activate repair mechanisms (Quesada et al. 1995). In Lake Biwa, higher UVB/UVA ratios have been correlated with cyanobacterial blooms, in particular toxic species of the genus *Microcystis*, which can alter the UV waveband ratios and make UVB exposure relatively more severe for other organisms in the water column (Frenette et al. 1996)

There is generally more energy available for photosynthesis in the littoral than in the pelagic zone in Lake Biwa. This greater average irradiance in the water column contributes to the higher plant productivity of the littoral which in turn is likely to favour production at higher trophic levels. Fluvial Lake Saint-Pierre is a shallow water ecosystem with adequate irradiance for macrophyte development throughout the benthic region of the lake. These shallow waters are biologically productive and support a large commercial perch fishery (Guénette et al. 1994). However, the greater UV penetration in the littoral zone of Lake Biwa and the South zone area in Lake Saint-Pierre also results in a higher potential for UV-induced damage of living organisms via a range of mechanisms (Vincent and Neale 2000). The aquatic biota living in the shallow littoral zone are less able to escape from UV exposure by moving out of the bright irradiance regime (Roy 2000) since the high UV-exposed environment is occupying a larger portion of the water column. However this continuous high exposure may select for UV-tolerant organisms and induce various cellular protection mechanisms (Roy 2000). Furthermore, UV may induce positive effects by increasing photodegradation of otherwise refractory CDOM, thereby enhancing bacterial growth by providing lower molecular weight carbon substrates (Waiser and Robarts 2000). These UV photochemical reactions may also increase UVR penetration into the water column by bleaching the CDOM (Gibson et al. 2000; Moran et al. 1999; Morris and Hargreaves 1997).

These results draw attention to the large microhabitat variability of the littoral zone and the inshore waters of the Lake Biwa and Lake Saint-Pierre ecosystems. This bio-optical variability of inshore waters should contribute to an increased physical complexity of the littoral zone, thereby favouring a broad range of species. Optical factors may also operate as selective pressures for organisms living in the littoral region of the lake due to the greater UV exposure and other changes in the underwater spectral composition. This may involve the disappearance of sensitive species or the selection for UV-resistant species. Differences in light climate are also likely to affect the nutritive value of phytoplankton which in turn affects organisms at higher levels in the aquatic food web. For example higher lipid and stoichiometric content have been found for biofilms growing in the South shore than the North shore of Lake Saint-Pierre (Huggins et al., in prep). As a result, the variability in light environment is likely to affect the biological communities, and may influence shifts in littoral biodiversity of the type reported in Lake Biwa (Nakanishi and Sekino 1996).

Bio-optical instrumentation for underwater applications now allows consideration of the local variability within the inshore waters of lakes. As shown here, *in situ* measurements such as spectral attenuation of UV and PAR (light available for photosynthesis) allow the rapid characterization of water masses. These measurements are important for assessing the inshore environment and offer promising tools for synoptic surveys and comparisons among many sites. These observations also draw attention to the large-scale, local variations in the littoral environment that must be addressed in any long term assessment of lake water quality.

Acknowledgments

This research was funded by a) the Lake Biwa Museum and Shiga prefectural office and, b) the Natural Sciences Research Council of Canada (NSERC) and the Fonds pour la formation des Chercheurs et l'Aide à la Recherche (FCAR) to J-J. F. We are very grateful to Dr. Yasushi Kusuoka, Prof. Hori from Kyoto University and his students and the Lake Biwa Museum staff for their invaluable support during the Lake Biwa study. We thank Louise Ferland and Hidetaka Yoneyama, for their strong logistic support and positive attitude throughout the Lake Biwa study. We thank Drs. Hiroki Haga, Kayoko Kameda, Kunihiko Kuwamura and Katsuki Nakai who supported the scientific work at various stages of the study. We thank Dr. Michio Kumagai for his encouragement of our optical research on Lake Biwa. We are very grateful to Christine Barnard, Hughes Boulanger, Claude Belzile, Julie Harnois, Kim Huggins, Marianne Lefebvre, Carl Martin, Olivier Mathieu, and Véronique Aubin for their invaluable help in the field and in the lab. Finally, we thank Captain Guy Morin and the scientific staff (Jean Morin and Daniel Rioux) of the ship "Le Pêcheur", of the Meteorological Service of Canada during the Lake Saint-Pierre study.

References

BELZILE, C., W. F. VINCENT, AND M. KUMAGAI. 2002. Contribution of absorption and scattering to the attenuation of UV and photosynthetically available radiation in Lake Biwa. Limnol. Oceanogr. **47:** 95-107.

EFFLER, S. W., M. G. PERKINS, AND D. L. JOHNSON. 1991. Optical heterogeneity in Lake Champlain. J. Great Lakes Res.**17:** 322-332.

FRENETTE, J.-J., S. NAKANO, T. NAKAJIMA, M. KUMAGAI, AND C. JIAO. 1996. An Enclosure Experiment in a Eutrophic Area in Lake Biwa - Cyanobacterial control of light and temperature during algal succession. Japanese Society of Limnology Meeting. -Hokkaido, September 25.

________, W. F. VINCENT, AND L. LEGENDRE. 1999. Size-dependent C:N uptake by phytoplankton as a function of irradiance: Ecological implications. Limnol. Oceanogr. **43:** 1362-1368.

________, M. T. ARTS, AND J. MORIN. 2002. Spectral gradients of downwelling light in a fluvial lake (Lac Saint-Pierre, St-Lawrence River). Aquatic ecology. In press.

GIBSON, J. A. E., W. F. VINCENT, AND R. PIENITZ. 2000. Hydrologic control and diurnal photobleaching of CDOM in a subarctic lake. Arch. Hydrobiol. **152:** 143-159.

GUÉNETTE, S., Y. MAILHOT, I. MCQUINN, P. LAMOUREUX, AND R. FORTIN. 1994. Paramètres biologiques, exploitation commerciale et modélisation de la Perchaude (*Perca flavescens*) du lac Saint-Pierre. Québec, Ministère de l'Environnement et de la Faune et Université du Québec à Montréal. 110p.

KIRK, J. T. O. 1994. Light and Photosynthesis in Aquatic Ecosystems. 2nd ed. Cambridge University Press, United Kingdom. 509p.

ISHIKAWA, K., M. KUMAGAI, W. F. VINCENT, S. TSUJIMURA, AND H. NAKAHARA. 2002. Transport and accumulation of bloom-forming cyanobacteria in a large, mid-latitude lake: the gyre-*Microcystis* hypothesis. Limnology **3**: 87-96.

LANGLOIS, C., L. LAPIERRE, M. LEVEILLE, P. TURGEON, AND C. MENARD. 1992. Synthèse des connaisances sur les communautés biologiques du Lac Saint-Pierre. Rapport technique. Zone d'intérêt prioritaire. Centre Saint-Laurent, Conservation et Protection, Environment Canada. 236p.

MORAN, M. A., W. M. SHELDON, AND J. E. SHELDON. 1999. Biodegradation of riverine organic carbon in five estuairies of the southeastern United states. Estuaries **22:** 55-64.

MORI, S., AND T. MIURA. 1990. List of plant and animal species living in Lake Biwa (corrected third edition).-Mem.Fac.Sci.Kyoto Univ. (Ser.B) **14:** 14-22.

MORRIS, D. P., AND R. B. HARGREAVES. 1997. The role of photochemical degradation of dissolved organic matter in regulating UV transparency of three lakes of the Plocono Plateau. Limnol. Oceanogr. **42:** 239-249.

NAKANISHI, M., AND T. SEKINO. 1996. Recent drastic changes in Lake Biwa bio-communities, with special attention to exploitation of the littoral zone. GeoJournal **40:** 63-67.

OLIVER, R. L. AND G. G. GANF. 2000. Freshwater blooms, p. 149-194. *In* B. A. Whitton, and M. Potts [eds.], The ecology of cyanobacteria. Kluwer Academic Publishers.

QUESADA, A., J-L. MOUGET, AND W. F. VINCENT. 1995. Growth of Antarctic cyanobacteria under ultraviolet radiation: UVA counteracts UVB radiation. J. Phycol. **31:** 242-248.

ROY, S. 2000. Strategies for the minimisation of UV-induced damage, p. 177-205. *In* S. J. De Mora, S. Demers and M. Vernet [eds.], The effects of UV radiation in the marine environment. Cambridge University Press, United Kingdom.

SEEHAUSEN, O., J. J. M. VAN ALPHEN, AND F. WITTE. 1997. Cichlid fish diversity threatened by eutrophication that curbs sexual selection. Science **277:** 1808-1811.

SMITH, R. E. H., J. A. FURGAL, M. N. CHARLTON, B. M. GREENBERG, V. HIRIART, AND C. MARWOOD. 1999. Attenuation of ultraviolet radiation in a large lake with low dissolved oxygen concentration. Can. J. Fish. Aquat. Sci. **56:** 1351-1361.

STERNER, R. W., J. ELSER, E. J. FEE, S. J. GUILDFORD, AND T. M. CHRZANOWSKI. 1999. Light:nutrient ratio in lakes: balance of energy and material affects ecosystem structure and process. Am. Nat. **150:** 663-684.

TSUDA, R., AND M. NAKANISHI. 1988. The relative importance of Chlorophyll *a* , non-living suspended and dissolved matter to the vertical light attenuation in the North Basin of Lake Biwa. Mem.Fac. Sci. Kyoto Univ. (Ser.Biol.) **13:** 101-109.

VINCENT, W. F., AND J. J. DODSON. 1999. The need for an ecosystem-level understanding of large rivers: the Saint Lawrence River, Canada-USA. Japn J. Limnol. **60:** 29-50.

________, AND P. J. NEALE 2000. Mechanisms of UV damage to aquatic organisms, p. 149-176. *In* S. J. De Mora, S. Demers and M. Vernet [eds.], The effects of UV radiation in the marine environment. Cambridge University Press, United Kingdom.

________, M. KUMAGAI, C. BELZILE, K. ISHIKAWA, AND K. HAYAKAWA. 2001. Effects of seston on ultraviolet attenuation in Lake Biwa. Limnol. **2:** 179-184.

WAINMAN, B. C., R. E. H. SMITH, H. RAI, AND J. A. FURGAL. 1999. Irradiance and lipid production in natural algal populations, Chapter 3, p. 45-70. *In* M. T. Arts and B.C. Wainman [eds.], Lipids in Freshwater Ecosystems. Springer-Verlag. New York.

WAISER, M. J., AND R. D. ROBARTS. 2000. Changes in composition and reactivity of allochthonous DOM in a prairie saline lake. Limnol. Oceanogr. **45:** 763--774.

WETZEL, R. G. 2001. Limnology. Academic press. San Diego.USA. 1006p.

2-3. Generic Approaches Towards Water Quality Monitoring Based on Paleolimnology

Reinhard Pienitz and Warwick F. Vincent

Centre d'études nordiques, Université Laval, Québec G1K 7P4, Canada

Abstract

Long term environmental records for lake and river ecosystems provide a valuable generic tool for water quality management. These data sets can play a pivotal role in determining natural baseline conditions, detecting early evidence of change, identifying the causal mechanisms of water quality deterioration, and in gauging the success of remediation measures. At most sites, however, such data are sparse or completely lacking. New advances in paleolimnology, that is the study of past environments based on the analysis of sediments, offer considerable potential for reconstructing these historical records. This paleolimnological approach is illustrated by way of water quality research on three ecosystems in Québec, Canada. Lake St-Augustin is a small lake characterized by episodes of bottom-water anoxia and summer blooms of cyanobacteria that result in its municipal closure to swimming and other lake activities for several weeks each summer. A paleolimnological analysis based on fossil diatoms showed that there have been four phases of nutrient enrichment over the last 240 years coinciding with initial colonisation and land development (1760-1900), farm development (1900-1950), increased fertiliser use and intensification of agriculture (1950-1980), and major road and residential expansion (1980-present). The paleolimnological application of diatom–based transfer functions for total phosphorus analysis of Lake St-Charles, the principal drinking water supply for Québec City, showed that substantial changes took place in the lake coincident with the raising of water level in the 1930s. There was no evidence of increasing eutrophication since that time, contrary to public perception. Finally, geochemical analysis of sediments in the St-Lawrence River showed greatly reduced concentrations of heavy metals and other pollutants over the period 1960-90, but the paleolimnological record also underscores the need for ongoing improvements in pollution control measures.

Introduction

The restoration and protection of freshwater ecosystems is an increasingly important priority for many environmental and conservation agencies throughout the world. A common problem in water quality management is the absence of reliable long term data series that provide information about the natural (pre-anthropogenic) "baseline" conditions in lakes and rivers that can be used to gauge the importance of measured or perceived changes in the present-day environment. For some freshwater issues it is important to detect early evidence of change and to know for how long that trend has persisted. This in turn raises questions such as whether any changes in key water quality variables are accelerating, whether these values have moved outside the bounds that were typical of the lake or river in the past, and whether there are natural cycles of variability to consider. For some issues it may be important to date precisely any past shifts in water quality and thereby identify causal changes in the surrounding catchments. Finally, for those lakes and rivers undergoing costly rehabilitation, it is important to determine how fast the waters are recovering, and to what extent the environment is returning to conditions prior to industrial, urban or rural development. Such information will help to guide ongoing treatment decisions and to set realistic legal or management goals (Rast and Holland 1988), as well as give important feedback to stakeholders including regulatory and funding agencies, politicians and the public. All of these questions require detailed long-term records, but in most cases such data are sparse or completely lacking.

Over the last two decades, paleolimnology, that is the study of past environments based on the analysis of lake sediments, has come to prominence with a suite of new high resolution sampling and dating protocols, new analytical approaches and refined statistical techniques for model building and application. Paleolimnological approaches have achieved success in helping to understand how lakes and their associated drainage basins respond to many types of environmental change including acidification, drought, temperature and other climatic shifts, nutrient enrichment and pollution by contaminants (reviewed in Smol 2002). Of special interest to the management of lake water quality, these techniques now allow a detailed, high-resolution record of environmental variables to be reconstructed on the basis of sediment cores obtained from lakes, ponds, reservoirs, wetlands, peatlands and large rivers. Paleolimnological analysis provides a compelling example of how a generic (global) approach can be applied to specific sites, although regional and local knowledge is still necessary for reliable interpretations.

In this chapter we describe the basic steps in paleolimnological analysis and then examine three case studies in the province of Québec, Canada, illustrating the types and utility of this approach. We first examine Lake St-Augustin where the appearance of cyanobacterial blooms has resulted in restrictions on the use of this lake for recreational purposes. Our second example is Lake St-Charles, drinking water supply for the city of Québec, where perceived changes in water quality as well as recent limnological measurements from the lake have created public concern. Our third example is the St-Lawrence River where an expensive program to identify and control industrial pollution has raised questions about the degree of recovery and the efficacy of ongoing control measures. Finally we conclude by summarizing some of the strengths, limitations and additional requirements of the paleolimnological approach to water quality monitoring and management.

Paleolimnological Analysis

The key steps in paleolimnological analysis are sediment coring, splitting and dating; geochemical and paleobiological analysis of the subsamples; and the development and application of transfer functions to reconstruct past environmental conditions (Fig. 1).

Sediment sampling and dating

Sediment coring is typically performed in the central, deepest water part of the lake to provide an overall integrative assessment of lake water conditions because this is where sediments from littoral and pelagic zones accumulate through sediment focusing processes (Blais and Kalff 1995). Cores taken from this accumulation zone within lake basins generally provide the most detailed (highest temporal resolution) and continuous paleolimnological records. However, local bays may have separate water quality problems (see the preceding chapter) and therefore require separate sampling and analysis. Deepwater sampling also minimizes wind-driven resuspension of sediments while maximizing the possibility of anoxic conditions that will tend to reduce effects of

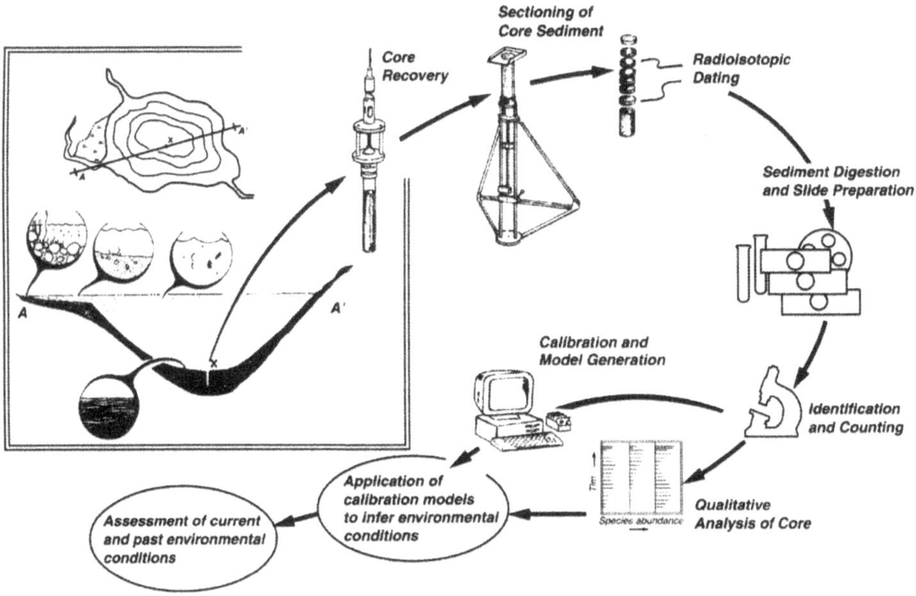

Fig. 1. Outline of the paleolimnological approach to environmental reconstruction (modified from Dixit et al. 1992).

bioturbation, that is the disruption of sediment profiles caused by the burrowing and feeding activity of benthic animals.

In most cases, lake sediments are deposited in a continuous fashion through time, with the most recent material overlying older sediments. Occasionally some problems in stratigraphic integrity occur (e.g., bioturbation, wind-induced turbulence), but these problems can often be identified and assessed. An undisturbed and continuous record of sediments can be retrieved from most lakes using a wide variety of coring techniques, usually including rod-driven piston corers (Livingstone-type), gravity corers (e.g., Kajak-Brinkhurst, Hongve, Glew, Limnos, and many others), or chamber-type samplers (e.g., freeze corers). The advantages and disadvantages of these different types of corers with respect to sediment core collection and extrusion are discussed in detail by Glew et al. (2001).

The sediment chronology is typically established using isotopes. The radioisotopic decay of the naturally occurring ^{210}Pb is now the method of choice to calculate dates of sediment layers over the recent past (up to about 150 years ago), whereas ^{14}C measurements can be used to estimate dates over the millennia. In some cases, a surprisingly high degree of temporal resolution (annual to sub-annual) can be attained in sediments with annual layers (couplets) called varves. A variety of other geochronological techniques are available including ^{137}Cs and ^{241}Am released by nuclear weapons testing and the nuclear industry (reviewed in Bradley 1999; Appleby 2001).

Sediment analysis – the importance of diatoms

Diatoms (class Bacillariophyta) are an important component of algal assemblages in lakes, comprising a large portion of total algal biomass over a broad spectrum of lake trophic status. They are valuable indicators for water quality monitoring (Stevenson and Smol 2003) and have been used extensively in paleolimnological assessments of changes in lake trophic state (reviewed in Hall and Smol 1999). Diatoms are excellent paleo-indicators because their siliceous cell walls (frustules) which can be identified to species level are generally abundant, diverse, and well preserved in lake sediments. It is common to find several hundred taxa in a single sediment sample, providing considerable ecological information concerning lake eutrophication. Diatoms also respond rapidly to eutrophication and recovery. Their rapid growth and immigration rates and the lack of physical dispersal barriers ensure there is little lag-time between perturbation and response, thereby making them early indicators of environmental change. Moreover, the environmental optima and tolerances of many taxa are generally well known; however, species may respond directly to variables related to lake trophic state, such as phosphorus and nitrogen, but also indirectly to related limnological variables (e.g., stratification patterns, water transparency and depth, ionic concentration and composition, and other water chemistry variables). A major challenge for the paleolimnologist is to determine which environmental variables are related to species assemblages, and to effectively use these inferences in a paleoenvironmental context.

Sediment analysis - other indicators

Given the increased relevance of paleolimnological studies to contemporary limnology and management, greater attention is now being given to organisms other than diatoms, in an effort to broaden the reconstructions to other trophic levels and communities. Holistic or multi-disciplinary approaches (e.g., diatoms, chrysophytes, algal pigments, zooplankton, benthic invertebrates, macrophytes [via plant macrofossils such as leaf tissue]) used in conjunction with transfer function techniques offer a broader perspective of the response of lakes and rivers to disturbance. For example, it is now possible to infer past fish population structure from aquatic insect and zooplankton remains (e.g., Uutala 1990; Jeppesen et al. 1996) and address the problem of historical trophic interactions. The application of the different indicators to the reconstruction of the trophic state of a lake is outlined in detail in Smol et al. (2001a,b).

In addition to the various biological indicators, a wide range of physical and geochemical techniques are available to study the inorganic fraction of lacustrine and riverine sediments (reviewed in Last and Smol 2001). The sedimentary (e.g., particle size, loss-on-ignition or organic matter content), magnetic (e.g., magnetic susceptibility) and geochemical properties (including elemental and isotopic geochemistry) of sediments can yield a wealth of information on the paleolimnology of freshwater ecosystems. For example, bulk organic carbon (^{13}C) and nitrogen (^{15}N) stable isotopes as well as elemental C and N have been employed by Köster et al. (submitted) to study the impacts of anthropogenic activities (including forest clearance and railroad construction in the watershed) on the nutrient balance of New England (USA) lakes. In addition, the isotopes of various metals, such as lead and mercury, can also be used to track pollution sources. The recent introduction of near-infrared spectrometry to paleolimnological studies has allowed for more extensive examination of the chemical and biological constituents of sediments (reviewed in Korsman et al. 2001).

Transfer functions

Paleolimnological interpretations of changes in the physical and chemical conditions of lakes can be made qualitatively and/or quantitatively. Although quantitative inferences are clearly preferred, there is little point in attempting to quantify a weak environment-species relationship. Qualitative interpretations, however, are still very important in many studies, and should not be dismissed as of secondary importance.

Quantitative interpretations are based on the present-day ecological characteristics of algal species (i.e., their optima and tolerances) that have been estimated from their contemporaneous distributions with respect to limnological variables. Inferring the value of, for example, past phosphorus concentrations from fossil diatom remains preserved in lake sediments involves a two-step process. First, the relationship between diatom distribution and contemporary phosphorus is modelled using surface sediment 'calibration' or 'training' sets that are the most powerful methods to determine these relationships (e.g., Birks 1998). The basic approach is to choose a suite of study lakes (typically between 50 to 80, but usually the more the better) that span the limnological gradients of interest. From each study lake, the biological assemblages (e.g., diatom valves) preserved in the surface sediments (usually top 0.5 to 1.0 cm, representing the last few years of sediment accumulation) are identified and enumerated. These sediments provide an integrated sample, in space and time, of the taxa that have accumulated over the previous few years. From each study lake, environmental data, consisting of present-day physical, chemical, and biological variables that are likely to be ecologically important, are also collected. Due to logistic constraints, these data are often based on "spot" measurements from a single visit to the study sites. A number of multivariate statistical techniques can then be used to determine which environmental variables are most highly related to the biological assemblages. Variables that account for a significant and independent proportion of the variation in the biological assemblages will normally result in strong reconstruction models (Birks 1998). From these data, an inference model or transfer function is constructed, which is then used in a second step to infer, with known errors, environmental variables of interest from the fossil diatom assemblages (Fig. 1).

In paleolimnology, weighted averaging (WA) has become the most popular method for solving these two steps. WA combines ecological plausibility with simplicity and empirical predictive power; it is based on the non-linear, unimodal model of species response which is observed for many diatom taxa, and has lower prediction errors in comparison with other techniques. Given its superior performance, WA has been used to derive phosphorus transfer functions for a number of regions in North America and Europe that have been used to generate quantitative historical total phosphorus (TP) reconstructions (e.g., Fritz et al. 1993; Hall and Smol 1999; Anderson and Odgaard 1994; Bennion et al. 1996; Köster et al. submitted). Pan and Stevenson (1996) developed a diatom-TP inference model for wetlands in western Kentucky (USA), whereas Christie and Smol (1993) produced a transfer function for estimating TN concentrations from diatoms in southeastern Ontario (Canada) lakes. More recently, weighted averaging partial least squares regression (WA-PLS; ter Braak and Juggins 1993; Bennion et al. 1996) has been developed to take into account the residual correlations in the species data due to the effects of unmodelled or nuisance variables, thereby improving the predictive power of the WA coefficients (optima and tolerances) derived from the training sets. Bennion et al. (1996) provide an application of WA-PLS to infer lake eutrophication in southeast England, and discuss the merits of this technique. A comprehensive review by Birks (1995) provides information concerning the basic biological and statistical requirements of quantitative reconstruction procedures, as well as methods for assessing their ecological and statistical performance.

It is important to point out that the environmental optima and tolerances that paleolimnologists are estimating for indicator organisms from surface sediment "calibration" or "training" sets represent empirical relationships. The latter document the distributions and abundances of indicators with respect to the particular environmental and ecological conditions that exist in the study region. Different ecosystems and study regions will have different selection pressures, and therefore the environmental optimum and tolerance determined for a given species in one specific study region will not necessarily be identical to the optimum and tolerance estimated for the same taxon in another calibration data set. Also, given the strong seasonality of planktonic diatoms (e.g., Reynolds 1984) as well as their environment, weighted averaging models based on samples at a single time of year may yield statistical relationships that do not reflect the actual conditions during growth. Paleolimnologists recently have made efforts to improve calibration data sets by comparing the performance of models developed for different seasons as well as for different ecological components (planktonic versus periphytic) of the diatom communities (e.g., Bradshaw et al. 2002; Köster et al. submitted).

Paleolimnology of a Eutrophic Lake

Lake St-Augustin is a small lake in southern Québec (area = 0.7 km^2; Z_{max} = 6.1 m; Z_{av} = 3.5 m) and a popular site for summer and weekend boating, swimming and other recreational activities. It lies some 20 km from the center of Québec City, but as the city has expanded this area has become a small but rapidly growing commuter satellite to the greater metropolitan region. This period of residential development also appears to have coincided with a rapid deterioration in water quality. During intermittent summer stratification the bottom waters of the lake become anoxic (Fig. 2), and blooms of cyanobacteria dominated by *Aphanizomenon* and *Microcystis* now regularly occur each summer and fall (M. Valentine Bouchard et al. unpublished data). These blooms in the years 2001 and 2002 resulted in the municipal closure of the lake to water contact sports and to fishing for several weeks. No long term environmental data are available for this lake, but it was of importance to know the history of deterioration of the lake and whether this was entirely a consequence of the expansion of the residential population.

A paleolimnological approach towards understanding the water quality trend in Lake St-Augustin was adopted by Roberge et al. (2002). A core was obtained from the lake at its deepest site and split into 0.5 cm subsamples. These were then dated by ^{210}Pb and analysed for their content and species composition of fossil diatoms. The surficial sediments were brown-olive in colour and flocculent, to a depth of 15 cm. At greater depth, the sediments were darker and more compact with a higher proportion of clay. There were conspicuous populations of red-pigmented oligochaetes, probably of the pollution-tolerant genus *Tubifex*, in the surficial 5 cm of sediment. The dating of the

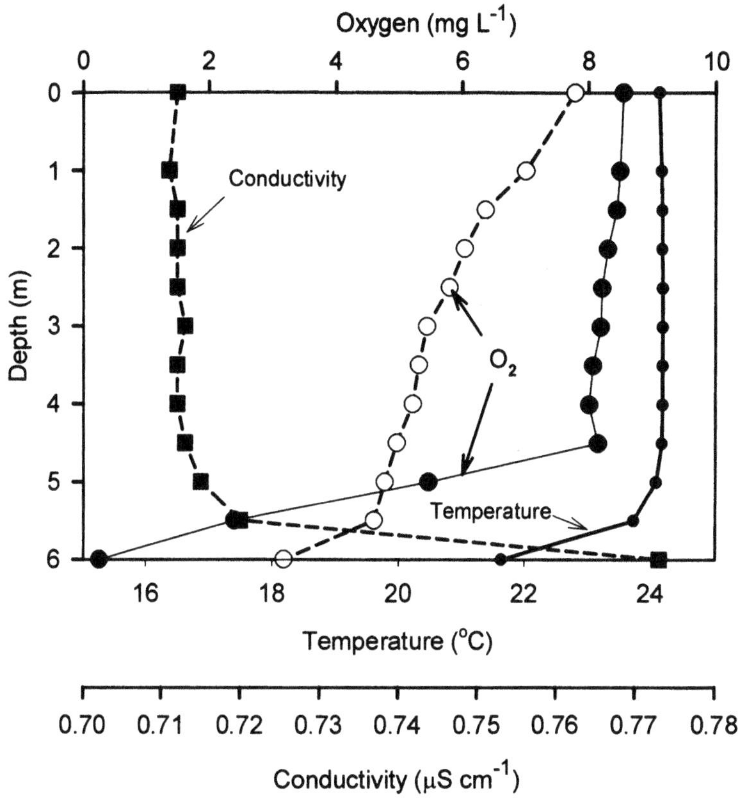

Fig. 2. Intermittent mixing, stratification and anoxia in Lake St-Augustin (Martin Bouchard Valentine et al. unpublished). The closed symbols are for 22 August 2002 (10h00) at the end of a period of stratification. Note the small temperature difference near the bottom of the profile and evidence from the conductivity profile of solute release from the sediments. The open symbols are for 23 August 2002 (10h00), for oxygen after storm-induced mixing.

core showed that the maximum depth of 31 cm corresponded to a date of 240 years before the present. The diatom analysis of the sediment core revealed 132 species, with substantial changes in community composition down the core indicating periods of major change in the nutrient status of the lake over the last two centuries (Fig. 3).

A period of slowly changing limnological conditions extended from 1760 (the date for the bottom of the core) to 1900. The community was dominated by centric diatom taxa such as the planktonic *Aulacoseira ambigua* and *Cyclotella pseudostelligera* (Fig. 3, zone 1) that indicated relatively good quality, oligo-mesotrophic conditions. There was a reduction over this period in some taxa such as the elongated form of

A. ambigua and slow rise in the proportional abundance of others such as the small benthic *Fragilaria pinnata*, suggesting the gradual onset of anthropogenic effects during this initial stage of human colonisation, forest use and land development (nearby Québec City was founded in 1608).

A substantial change took place over the period 1900-1950 (Fig. 3, zone 2), a period of major decline in *A. ambigua* and *Cyclotella* spp., sustained importance of *F. pinnata,* and the appearance of *Stephanodiscus hantzschii*, a small centric diatom species that is indicative of strongly polluted conditions. This would have corresponded to a period of major farm expansion and agricultural development of the catchment. A third abrupt change characterized the transition to post-World War II conditions (late 1940s to late 1970s; Fig. 3, zone 3). At this time the new species that had arrived in the previous period such as *Asterionella formosa* and *Fragilaria crotonensis* became important subdominants, and S. *hantzschii* accounted for up to 45% of the total counts. These observations imply greatly accelerated enrichment, and correspond to the post-war intensification of agriculture and the massive increase in fertiliser application that occurred in many parts of the world at that time.

The final stage in the sediment record is for the period late 1970s to the present and shows co-dominance by the pollution-tolerant taxa *F. crotonensis* and S. *hantzschii* (Fig. 3, zone 4). This period corresponds to the major expansion of residential developments and road construction within the lake's catchment, including a major highway. The increased occurrence of (exotic) salt-tolerant species in recent sediments reflects the inwash of salts used for the de-icing of roads during the winter months. These observations show that the degradation of Lake St-Augustin is the result of a long history of anthropogenic impacts, and that the ultimate restoration of this waterbody must take a similarly long-term perspective. In future paleolimnological studies of this lake it will be of great interest to identify the period of onset of cyanobacterial blooms given that these are the primary water quality concern. One approach towards this is the use of pigment markers in the sediments, as applied, for example, to lakes in the Canadian Prairies (Hall et al. 1999).

Paleolimnology of a Drinking Water Reservoir

Lake St-Charles lies within hilly terrain, about 20 km north of the city centre of Québec City. It is partially surrounded by residential and holiday homes, with 53 % of the catchment still in forest. From 1847 onwards the lake has been used as the principal drinking water supply for the city. Limnological analyses by our research group in the

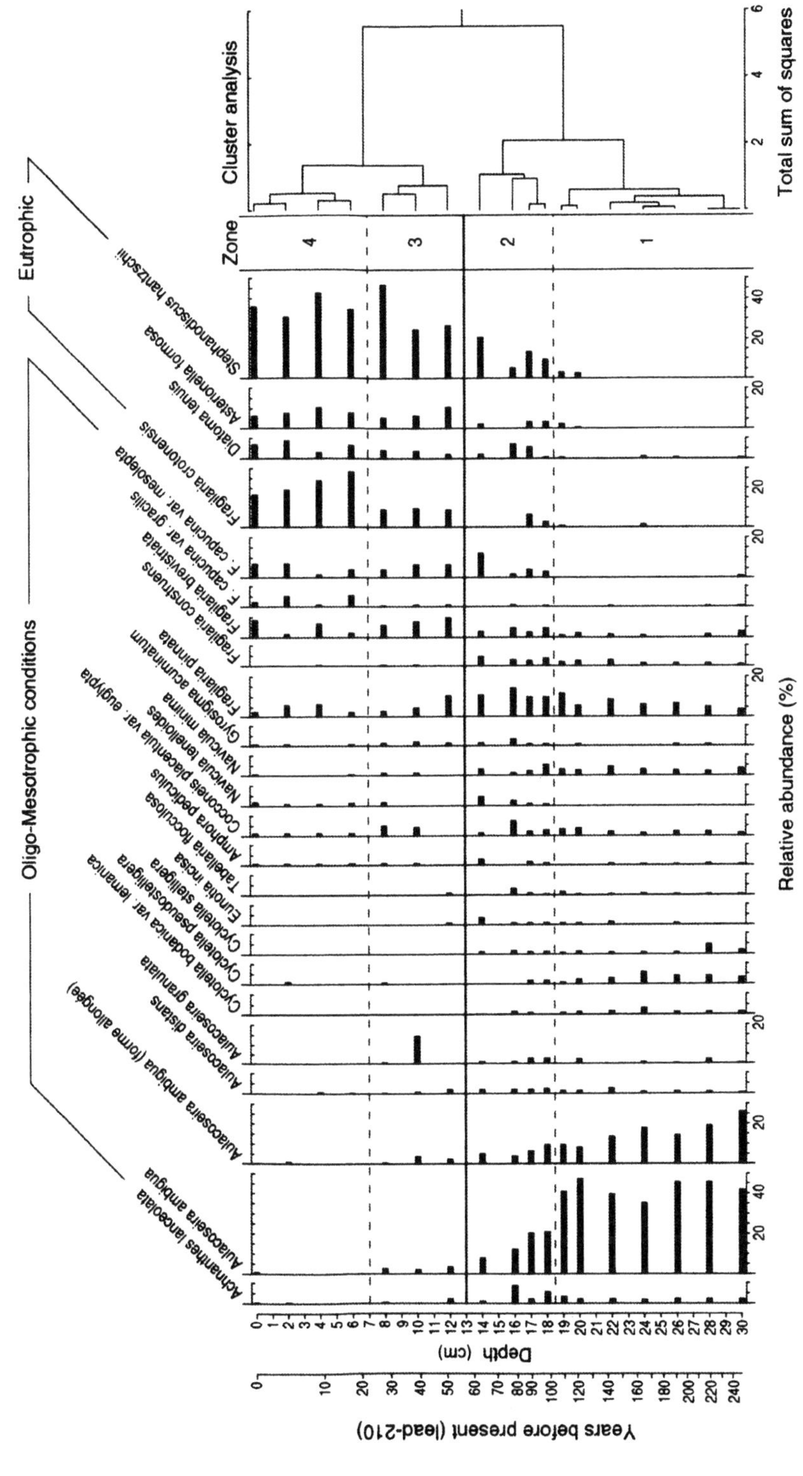

Fig. 3. Diatom analysis of a dated sediment core from Lake St-Augustin (modified from Roberge et al. 2002).

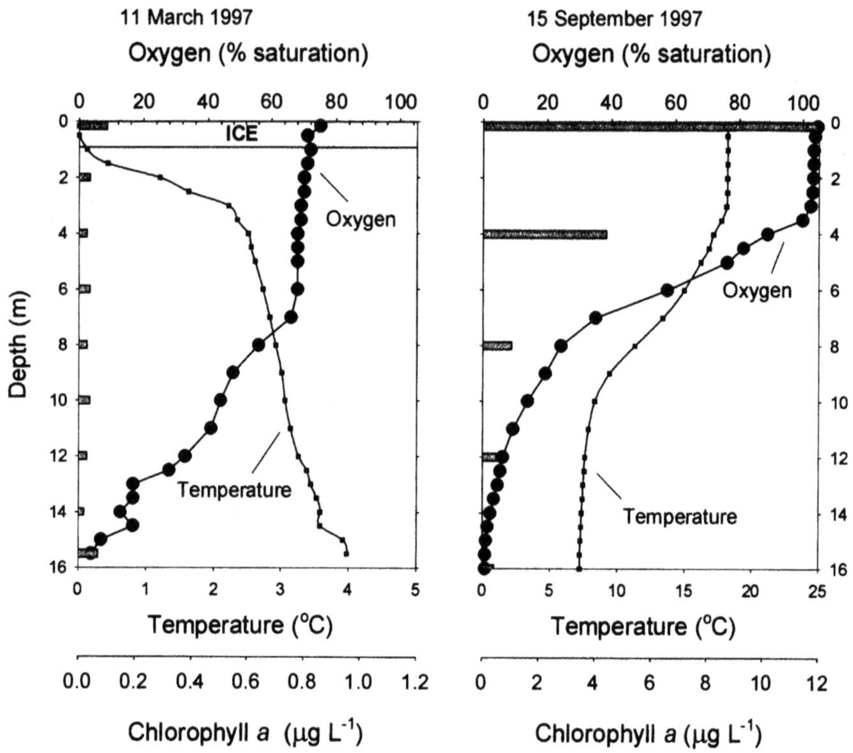

Fig. 4. Water column profiles in Lake St-Charles at the end of winter stratification under the ice, and at the end of summer stratification (modified from Tremblay et al. 2001). The bars represent Chlorophyll a concentrations.

late 1990s showed that this reservoir was showing signs of enrichment, with complete depletion of bottom water oxygen (anoxia) in the deepest part of the lake in late summer and late winter (Fig. 4). Moreover, some local residents believed that the lake was rapidly deteriorating in quality.

In order to estimate the extent of recent changes in trophic status and to identify critical periods of past anthropogenic disturbances, a paleolimnological analysis of Lake St-Charles sediments was undertaken in 1997 by Tremblay et al. (2001). Quantitative estimates of past total phosphorus (TP) concentrations in the water column of Lake St-Charles were obtained by applying a diatom-TP reconstruction model developed for 54 lakes located in south-eastern Ontario (Canada) on fossil diatom assemblages from a 28 cm long sediment core. The timing of changes in the fossil diatom record was estimated by ^{210}Pb dating. The study revealed changes in fossil diatom assemblage composition during the past ca. 150 years (Fig. 5), with the most striking biological and physico-

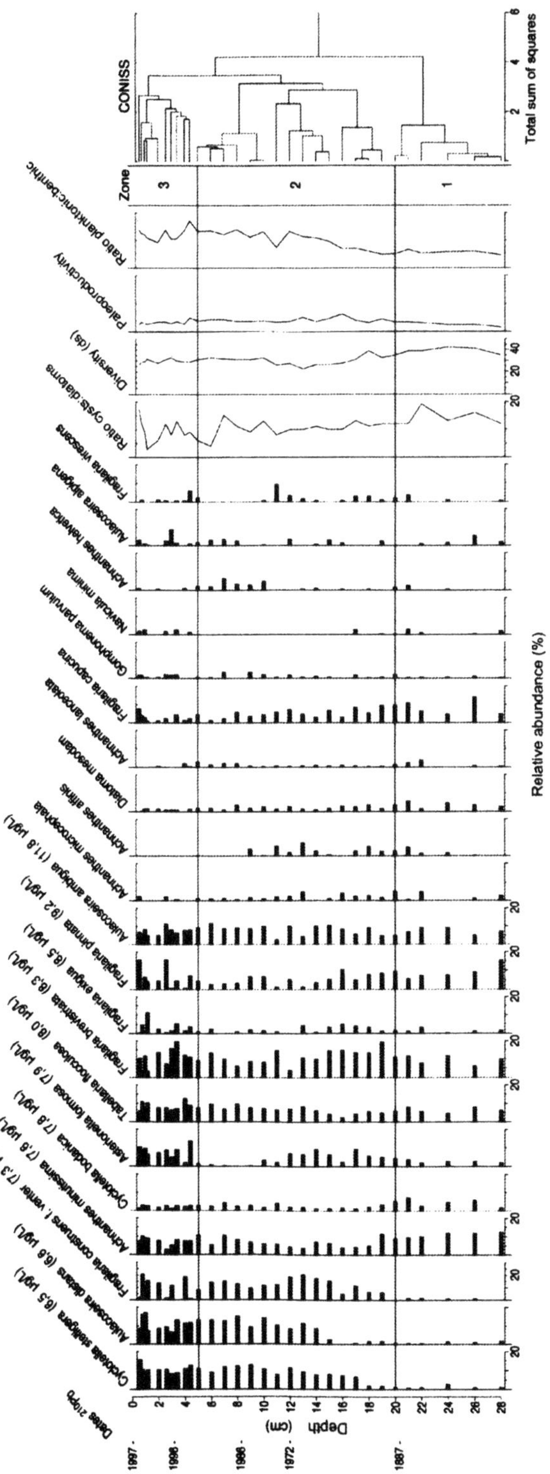

Fig. 5. Diatom analysis of a dated sediment core from Lake St-Charles (modified from Tremblay et al. 2001).

chemical changes occurring immediately after 1934. This date coincides with the construction of a dam, which raised the lake water level by 1.5-2 m. This modification was accompanied by significant shifts in diatom community structure, especially in the planktonic/benthic ratio (with increases in planktonic diatoms *Cyclotella stelligera* and *Aulacoseira distans*), and by changes in the physico-chemical characteristics of the sediments. Paleoproductivity increased at the same time, but remained more or less stable following conservation efforts between 1950 and 1970 (e.g., construction of a sewage treatment system).

The fossil diatom community structure indicates that mesotrophic conditions prevailed during the recent history of Lake St-Charles, and that diatoms typical of eutrophic conditions never became established in the lake. The diatom-inferred quantitative reconstruction of lake water total phosphorus (Fig. 6) revealed a slight decrease in total phosphorus over time, from close to 17 $\mu g \cdot L^{-1}$ prior to 1887 to about 13 $\mu g \cdot L^{-1}$ in recent times.

The fossil diatom analyses indicate that Lake St-Charles has not experienced significant recent changes in trophic status due to increased human activities in its drainage basin. However, our geochemical analyses show a sharp rise in metal concentrations (especially Fe, Mn, Cu, Pb and Zn), beginning in the late 19th century, reaching a plateau by the late 1970s (Fig. 7), which may be attributed to increased atmospheric pollution since the beginning of intense human colonization in the lake's catchment and surrounding areas. This in combination with the advanced mesotrophic status of the lake indicates the ongoing need for careful management of the watershed to prevent further changes in this important urban freshwater resource.

It will of interest in the future to determine at what point in time the bottom waters of Lake St-Charles became anoxic. One approach with considerable promise is the use of chironomids as paleo-indicator organisms. Some of these species have relatively narrow tolerances and require oxygenated conditions while others can tolerate low oxygen and anoxic conditions. Moreover, the head capsules of these insect larvae are relatively resistant to decomposition and therefore remain well preserved in the sediments (see Smol 2002). This approach has been applied with success to lakes in the Canadian prairies where the results showed that anoxic bottom-water conditions occurred well before the arrival of European settlers, and that the lakes were naturally eutrophic (Hall et al. 1999).

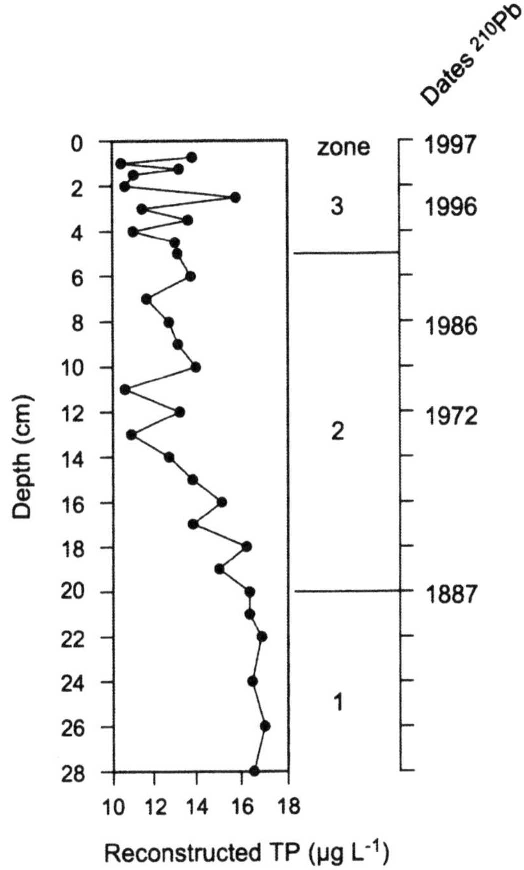

Fig. 6. Reconstructed total phosphorus concentrations in the surface waters of Lake St-Charles, determined from the application of a TP transfer function to the sediment diatom record (modified from Tremblay et al. 2001).

Paleolimnology of a Large River Ecosystem

The St-Lawrence River runs 500 km from Lake Ontario to the sea and is a major resource for navigation, industry, and agriculture. Additionally, it is the drinking water supply for almost half the population of the province of Québec, and is a rich ecosystem with a diversity of wildlife habitats (Vincent and Dodson 1999 and references therein). The river has been severely impacted over the course of the 20th century by industrial and other human activities. Remediation work began in the 1970s, and in the 1980s and 90s major efforts were undertaken to curb the pollution by major industries. Are these efforts resulting in improvement, and how far is the present-day environment from

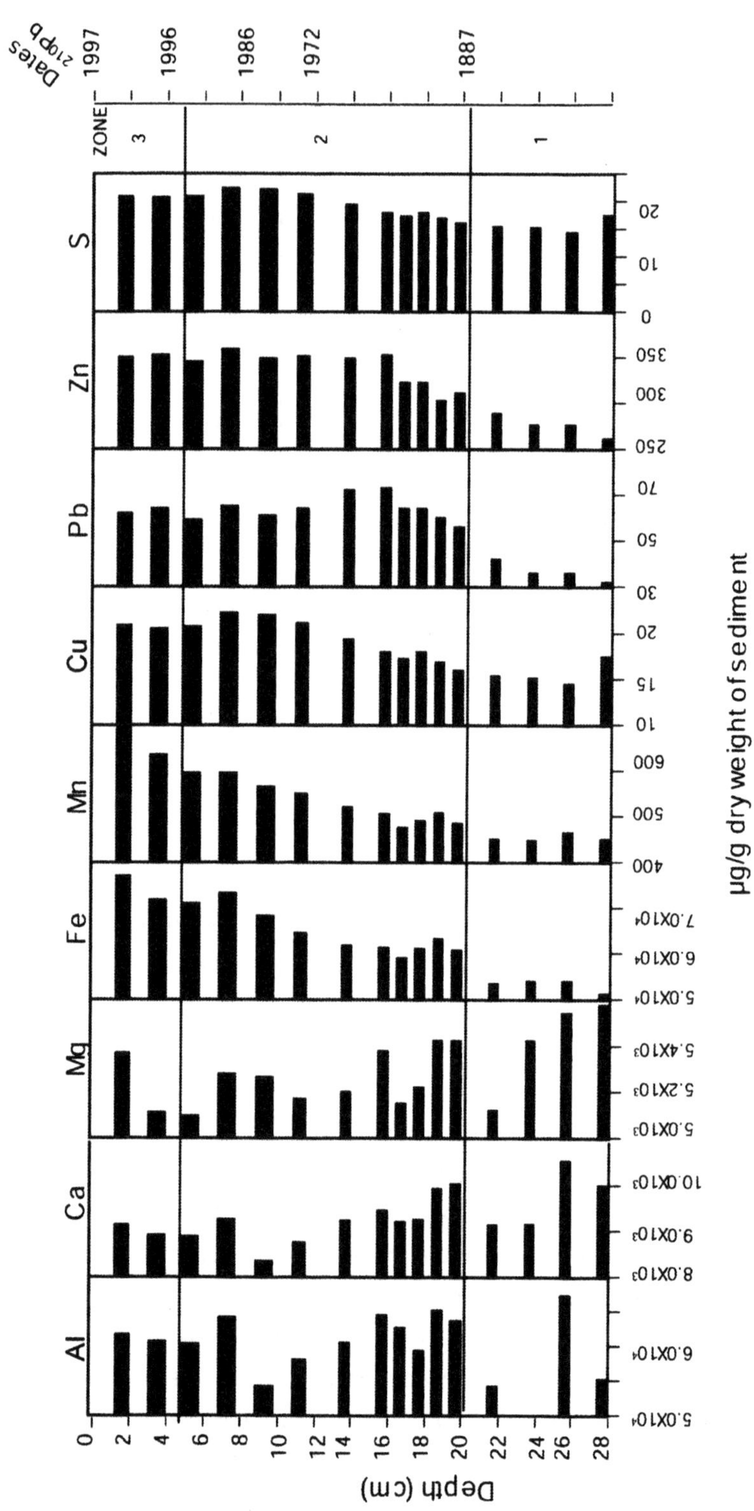

Fig. 7. Concentration of metals and sulfur in the dated sediment core from Lake St-Charles (modified from Tremblay et al. 2001).

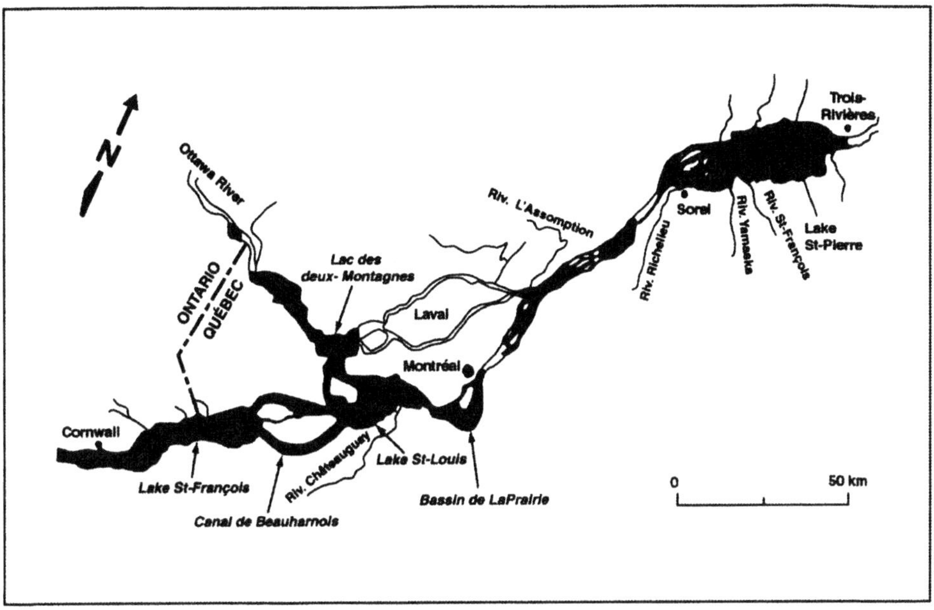

Fig. 8. The fluvial lakes of the St Lawrence River.

natural baseline conditions?

A problem in applying paleolimnological methods to river ecosystems is that fast flowing currents preclude the deposition of sediments, and flood events can completely scour the stream or river bed (Smol 2002). Furthermore, rivers tend to be well-oxygenated and large benthic populations of animals can result in substantial bioturbation. Despite these limitations, sediment cores have been successfully obtained from several parts of the St-Lawrence River, especially the more slowly flowing fluvial lakes (Fig. 8).

A detailed geochemical analysis was undertaken of a ^{210}Pb-dated sediment core from Lake St-Louis, immediately upstream of Montréal (Table 1; Carignan et al. 1994). This analysis showed how very high concentrations of organic contaminants were present in the river in the late 1950s and 60s, and dropped by a factor of 5-10 by 1990. These encouraging signs of improved water quality are also evident from the analysis of trace metals, with substantial reductions in cadmium, copper, lead and zinc. On the other hand, changes in chromium and nickel levels were relatively minor. Furthermore, concentrations of all metals still lay 2-7 times above the background values from pre-industrial strata indicating considerable room for ongoing improvement. These analyses

Table 1. Contamination of sediments in Lake Saint-Louis, St-Lawrence River. The data are compiled from the text and Fig. 6 of Carignan et al. (1994). Pre-industrial values are for the bottom 35-45 cm of a sediment core, > 130 years before present, containing 1.1% organic carbon. The metal values are in ppm and the organic contaminant values are in ppb.

Contaminant	Pre-industrial	Max. during the 1960s	1990
Trace metals			
Cd	0.15	3.5	1
Cr	62	140	120
Cu	17	70	48
Ni	32	70	60
Pb	15	72	38
Zn	78	750	260
Organic contaminants			
PCB #118	0	10	2
Mirex	0	0.8	0.1
DDD	0	8	0.2

also draw attention to the dangers of dredging sections of the St- Lawrence River which could result in resuspension and remobilization of sediments that were heavily contaminated in the middle of last century. Additional paleolimnological work has been undertaken in the St-Lawrence fluvial lakes to reconstruct changes in the riverine algal communities and to examine changes in macrophyte (water weed) biomass (Reavie et al. 1998). The results from Lake St-François sediments indicated a marked shift towards high macrophyte populations from the 1930s onwards. Eutrophic diatom taxa were present at their highest proportional abundance during the middle of last century, and then showed some decline in importance up to the end of the record (1990) indicating a recent improvement in water quality and a positive response to rehabilitation and control strategies. There were substantial differences between the paleolimnological records from Lake St-François and Lake St-Louis, indicating the heterogeneous nature of large river ecosystems and the need to consider local variability.

Conclusions

In this chapter, we have provided some of the many examples of how paleolimnological studies based on fossil diatoms and geochemical analysis can be used as powerful tools for eutrophication research and management. The historical perspective allows the assessment of natural variability and the establishment of baseline or pre-disturbance water chemistry conditions as targets for lake rehabilitation. The past decade was largely devoted to developing and refining diatoms as quantitative bioindicators of lake eutrophication. Current research includes assessing how a detailed knowledge of local conditions, for example the seasonal dynamics of the biota and their limnological environment, can be used to improve the application of transfer functions and further refine the interpretation of quantitative paleolimnological records.

Studies of diatoms alone do not allow a full assessment of complex food-web interactions in lakes and rivers, but fortunately there are many other aquatic and terrestrial biota that leave fossil remains in lake sediments (Smol 2002). For example, chironomids have been used to monitor deep-water oxygen levels (Quinlan et al. 1998), and in conjunction with diatoms they will provide a better understanding of the relationships between changes in upper and lower strata of the water column during eutrophication. Fossil pigments (including the remains of nitrogen-fixing cyanobacteria) and diatoms provide a useful measure of the biomass of all major algal groups, as well as total algal biomass (Leavitt and Hodgson, 2001). The combined use of biogenic silica analysis (Conley and Schelske 2001), diatoms, and fossil pigments may help clarify the relative roles of Si, P and N limitation during eutrophication. The combined application of these various bio-indicators and geochemical methods in multi-disciplinary studies will greatly strengthen our ecological understanding of lake eutrophication processes, as well as provide a set of powerful tools for water quality monitoring and management.

Acknowledgements

We thank Dr. Michio Kumagai for inviting us to contribute this article; the Natural Sciences and Engineering Research Council of Canada and Fonds pour la Formation de chercheurs et l'aide à la recherche for funding; and Centre d'études nordiques for logistic support.

References

ANDERSON, N. J., AND B. V. ODGAARD. 1994. Recent palaeolimnology of three shallow Danish lakes. Hydrobiologia **275/276:** 411-422.

APPLEBY, P. G. 2001. Chronostratigraphic techniques in recent sediments, p. 171-203. *In* W. M. Last and J. P. Smol [eds.], Tracking Environmental Change Using Lake Sediments, vol. 1, Basin Analysis, Coring, and Chronological Techniques. Kluwer Academic Publishers, Dordrecht.

BENNION, H., S. JUGGINS, AND N. J. ANDERSON. 1996. Predicting epilimnetic phosphorus concentrations using an improved diatom-based transfer function and its application to lake management. Env. Sci. Technol. **30:** 2004-2007.

BIRKS, H. J. B. 1995. Quantitative palaeoenvironmental reconstructions, p. 161-254. *In* D. Maddy and J. S. Brew [eds.], Statistical Modelling of Quaternary Science data, Technical Guide 5. Cambridge University Press, United Kingdom.

_________. 1998. Numerical tools in palaeolimnology – Progress, potentialities, and problems. J. Paleolimnol. **20:** 307-332.

BLAIS, J. M., AND J. KALFF. 1995. The influence of lake morphology on sediment focusing. Limnol. Oceanogr. **40:** 582-588.

BRADLEY, R. S. 1999. Paleoclimatology. Academic Press, San Diego, USA, 613p.

BRADSHAW, E. G., N. J. ANDERSON, J. P. JENSEN, AND E. JEPPESEN. 2002. Phosphorus dynamics in Danish lakes and the implications for diatom ecology and palaeoecology. Freshwat. Biol. **47:** 1963-1975.

CARIGNAN, R., S. LORRAIN, AND K. LUM. 1994. A 50-year record of pollution by nutrients, trace metals and organic chemicals in the St-Lawrence River. Can. J. Fish. Aquat. Sci. **51:** 1088-1100.

CHRISTIE, C. E., AND J. P. SMOL. 1993. Diatom assemblages as indicators of lake trophic status in southeastern Ontario lakes. J. Phycol. **29:** 575-586.

CONLEY, D. J., AND C. L. SCHELSKE. 2001. Biogenic Silica, p. 281-293. *In* J. P. Smol, H. J. B. Birks and W. M. Last [eds.], Tracking Environmental Change Using Lake Sediments, vol. 3, Terrestrial, Algal, and Siliceous Indicators. Kluwer Academic Publishers, Dordrecht.

DIXIT, S. S., J. P. SMOL, J. C. KINGSTON, AND D. F. CHARLES. 1992. Diatoms: Powerful indicators of environmental change. Environ. Sci. Technol. **26:** 22-33.

FRITZ, S. C., J. C. KINGSTON, AND D. R. ENGSTROM. 1993. Quantitative trophic reconstruction from sedimentary diatom assemblages: a cautionary tale. Freshwat. Biol. **30:** 1-23.

GLEW, J. R., J. P. SMOL, AND W. M. LAST. 2001. Sediment core collection and extrusion, p. 73-105. *In* W. M. Last and J. P. Smol [eds.], Tracking Environmental Change Using Lake Sediments, vol. 1, Basin Analysis, Coring, and Chronological Techniques. Kluwer Academic Publishers, Dordrecht.

HALL, R. I., AND J. P. SMOL. 1999. Diatoms as indicators of lake eutrophication, p. 128-168. *In* E. F. Stoermer and J. P. Smol [eds.], The Diatoms: Applications for the Environmental and Earth Sciences. Cambridge University Press.

_________, P. R. LEAVITT, R. QUINLIN, A. DIXIT, AND J. P. SMOL. 1999. Effects of agriculture, urbanization, and climate on water quality on the Northern Great Plains. Limnol. Oeanogr. **44:** 739-756.

JEPPESEN, E., E. A. MADSEN, J. P. JENSEN, AND N. J. ANDERSON. 1996. Reconstructing the past density of planktivorous fish and trophic structure from sedimentary zooplankton fossils: a surface sediment calibration data set from shallow lakes. Freshwat. Biol. **35:** 115-127.

KORSMAN, T., I. RENBERG, E. DÅBAKK, AND M. B. NILSSON. 2001. Near-infrared spectrometry (NIRS) in palaeolimnology, p. 299-317. *In* W. M. Last and J. P. Smol [eds.], Tracking Environmental Change Using Lake Sediments, vol. 2, Physical and Geochemical Methods. Kluwer Academic Publishers, Dordrecht.

KÖSTER, D., R. PIENITZ, B. WOLFE, S. BARRY, D. FOSTER, AND S. DIXIT. 2003. Human-induced impacts on the nutrient balance of Walden Pond during the last three centuries, as inferred by diatoms and stable isotopes. Can. J. Fish. Aquat. Sci. (submitted).

LAST, W. M., AND J. P. SMOL [eds.]. 2001. Tracking Environmental Change Using Lake Sediments, vol. 2, Physical and Geochemical Methods. Kluwer Academic Publishers, Dordrecht, 504p.

LEAVITT, P. R., AND D. A. HODGSON. 2001. Sedimentary pigments, p. 295-325. *In* W. M. Last, and J. P. Smol [eds.], Tracking Environmental Change Using Lake Sediments, vol. 2, Physical and Geochemical Methods. Kluwer Academic Publishers, Dordrecht.

PAN, Y., AND R. J. STEVENSON. 1996. Gradient analysis of diatom assemblages in western Kentucky wetlands. J. Phycol. **32:** 222-232.

QUINLAN, R., J. P. SMOL, AND R. I. HALL. 1998. Quantitative inferences of past hypolimnetic anoxia in south-central Ontario lakes using fossil midges (Diptera: Chironomidae). Can. J. Fish. Aquat. Sci. **55:** 587-596.

RAST, W., AND M. HOLLAND. 1988. Eutrophication of lakes and reservoirs: a framework for making management decisions. AMBIO **17:** 2-12.

REAVIE, E. D., J. P. SMOL, R. CARIGNAN, AND S. LORRAIN. 1998. Paleolimnological reconstruction of two fluvial lakes in the St Lawrence River: A reconstruction of environmental changes during the last century. J. Phycol. **34:** 446-56.

REYNOLDS, C. S. 1984. The Ecology of Freshwater Phytoplankton. Cambridge University Press, Cambridge.

ROBERGE, K., R. PIENITZ, AND S. ARSENAULT. 2002. Eutrophisation rapide du lac Saint-Augustin, Québec: étude paléolimnologique pour une reconstitution de la qualité de l'eau. Le Naturaliste Canadien **126:** 68-82.

SMOL, J. P. 2002. Pollution of Lakes and Rivers: A Paleoenvironmental Perspective. Oxford University Press, USA. 280p.

_________, H. J. B. BIRKS, AND W. M. LAST [eds.], 2001a. Tracking Environmental Change Using Lake Sediments, vol. 3, Terrestrial, Algal, and Siliceous Indicators. Kluwer Academic Publishers, Dordrecht, 371p.

_________, _________, AND _________ [eds.], 2001b. Tracking Environmental Change Using Lake Sediments, vol. 4, Zoological Indicators. Kluwer Academic Publishers, Dordrecht, 217p.

STEVENSON, R. J., AND J. P. SMOL. 2003. Use of algae in environmental assessments, p. 775-804. *In* J. D. Wehr and R. G. Sheath [eds.], Freshwater Algae of North America. Ecology and Classification. Academic Press, USA.

TER BRAAK, C. J. F., AND S. JUGGINS. 1993. Weighted averaging partial least squares regression (WA-PLS): an improved method for reconstructing environmental variables from species assemblages. Hydrobiologia **269/270:** 485-502.

TREMBLAY, R., S. LÉGARÉ, R. PIENITZ, W. F. VINCENT, AND R. HALL. 2001. Étude paléolimnologique de l'histoire trophique du lac Saint-Charles, réservoir d'eau potable de la communauté urbaine de Québec. Revue des Sciences de l'Eau **14:** 489-510.

UUTALA, A. J. 1990. *Chaoborus* (Diptera : Chaoboridae) mandibles- paleolimnological indicators of the historical status of fish populations in acid-sensitive lakes. J. Paleolimnol. **4:** 139-151.

VINCENT, W. F., AND J. J. DODSON. 1999. The need for an ecosystem-level understanding of large rivers: the Saint-Lawrence River, Canada-USA. Japanese J. Limnol. **60:** 29-50.

Chapter 3

Approaches Towards Environmental Restoration

3-1. Global Perspectives and Limitations of Lake Restoration

Tom Murphy

National Water Research Institute, Ecosystem Science Directorate,

Environmental Conservation Service, Environment, Canada

Abstract

Some lake restoration problems are clearly global, and local treatments have little chance of success. Introduced species are one example where prevention has to be considered as the primary treatment. In the Great Lakes, introduced species like zebra mussels and round gobies are changing the ecosystem and threatening native species with extinction. There has been some control of introduced sea lamprey but only at great expense. Certainly, atmospheric loading of toxins like PCBs and dioxins into lakes must also be controlled at the source. Other contaminants, like mercury, probably need the same management, but consensus is not as clear. Once persistent contaminants are released into large bodies of water, they cannot be effectively treated. Sulfur loading is another aspect of atmospheric loading that is best controlled at the source. Sulfur loading into Lake Biwa, Japan is causing enhanced eutrophication. It is better to help continental Asia control their sulfur releases than to try to treat the problem in Lake Biwa. When lake restoration is used at a local level, it is still important to have at least a regional or national coordination. Treatment of lakes is often too expensive for local taxes and a higher level of government must be involved. For the most serious global water management problems, international cooperation is required. Water deficits in some areas are serious and may involve climate change driven by global factors like increases in carbon dioxide. Water management in the third world often requires global cooperation.

Introduction

The global problems in water management require at least improved implementation of current technologies. However some problems require new management procedures and changes in lifestyle. Many times eloquent experts point at the foolishness of using water on lawns but most suburbs in North America remain the same. At capital costs estimated to be up to $6.2 billion, the USA is about to remove most arsenic from drinking water (Reid 1994, Smedley and Kinniburgh 2002). Water must be priced so that it is used effectively. Although technical people usually understand the significance of how water sources are changing, at times many people do not. Moreover, certain industries are hesitant to accept concepts like global warming. As the world population becomes denser and more industrialized, the concept of using dilution as a solution to pollution continues to weaken.

The effects of global warming on the world ecosystems are "clearly visible" (Walther et al. 2002) and will likely result in increased extinction of species with little tolerance and diminishing habitat. I will focus on other issues that are less well reported but certainly related to global warming.

Limitations on Availability of Fresh Water

One of the "simplest" aspects of global change is the lack of water. Population growth is partly responsible, but as some cultures become industrialized, their water requirements increase much more than population would indicate. For example, because of water extraction, groundwater in northern China is dropping greatly, and for much of the year there is no flow in major rivers like the Yellow River. More conservative use of water is mandatory but alternative sources of water are usually sought first.

Increased use of groundwater produces a number of problems. Land subsidence is the most well known. Some groundwater is toxic, and it seems that increased use of groundwater can contribute to its toxicity. Excessive use of groundwater may result in oxidation and the mobilization of arsenic. Millions of villagers in Bangladesh are being poisoned by arsenic in groundwater (Karim 2000). If groundwater is still to be used, it must be recharged during the monsoons. Groundwater geochemistry must be managed to ensure that only good water is used.

Massive diversions of water from the Yangtze River are underway to take water from the south to the north of China. In Russia, so much water is diverted for irrigation that the Aral Sea is disappearing. Transboundary movement of water is more difficult. In North America, there are many discussions of diverting water from Canada to the USA. Canadians are very sensitive about the changes in lifestyle that could occur if water was diverted. Residents of Oregon and Washington are equally as sensitive about ideas of diverting the Columbia River south to California. Most other water diversions would involve negotiations between countries with less similarity of cultures than in North America, and the potential for war over such diversions is serious. The politics, economics and physical limitations will likely prevent many diversions. Usually the effects of diversions and reservoirs seem localized but they are not. The changes in water storage in the northern hemisphere are so large that there has been a polar drift of the earth's rotational axis over the last 40 years (Chao 1995). The effects of such water movement and storage on global climate change are likely less than those induced by changes in carbon dioxide emissions, but they would have many localized effects. A more obvious effect of water movement is the enhanced homogenization of fish assemblages and loss of regional distinctness (Scott and Helfman 2002).

Technology is allowing water movement to occur in manners that had been impossible a decade ago. The desalinization of saltwater is resulting in unusual water discharge problems. The discharge of hypersaline water from desalinization plants on the coast must also be managed to prevent toxicity to corals and other marine life. The increased use of desalinization in the Persian Gulf results in increased discharge of wastewater to aquifers. The upwelling of groundwater in urban areas has concerned managers. The exposed water may create breeding habitat for mosquitoes and result in the spread of malaria that now is only present in the neighboring hills. Appropriate plumbing should resolve many concerns but major changes in water management in the Persian Gulf could have surprising consequences. The proposed pumping of desalinated water into aquifers for storage in an area with well-known major geological faults has the potential to induce an earthquake. If a quake broke major oil pipelines, the effect would be felt around the globe. In most of the world, geotechnical instability induced by water management is more commonly associated with dams.

Another unusual management issue of water in the Persian Gulf is the effect of pumping water down offshore wells to enhance extraction of oil. One concern is the potential for mobilization of toxic brines that can be associated with oil and gas. In many parts of the world such brines are enriched with mercury and arsenic. In Alberta, Canada, the pumping of freshwater to enhance recovery of oil has recently upset citizens who believe that such water use is resulting in droughts (Nikiforuk 2002). The effects of such pumping upon sulfur rich brines are very contentious with some Albertans complaining about poor air quality from reduced sulfur. The odor problem might be mitigated with appropriate monitoring and subsequent local treatment, but complete resolution would only come with use of other energy sources.

Shipping

Another physical aspect of water that causes major ecosystem disruptions is use of water in ship ballasts. For stability, big ships need to pump water into empty holds. Unfortunately, they usually discharge this water with its associated animals, microbes and plants in ports. One of the first introduced species in the Great Lakes was the sea lamprey. It greatly suppressed commercial fisheries and has been controlled at great cost and with many technologies. Use of infertile males and traps is relatively innocent but earlier, toxins were commonly used. More recently in the Great Lakes, introduced zebra mussels (*Dreissena polymorpha*) and quagga mussels (*Dreissena bugensis*) have dramatically changed the Great Lakes (Ricciardi 2001). These mussels increase light penetration and appear to stimulate growth of toxic algae. However, many variables have changed at once, so the analysis is complex. Zebra mussels ingest toxic algae but reject them in pseudofeces. The deposits of pseudofeces are prone to resuspension and presumably during storms enter the water column. It is well known that storms are commonly associated with kills of fish and birds, but storms also can kill fish via thermal shock. Also, the enhanced light generated by zebra mussel feeding allows macrophytes to grow to greater water depths. When storms occur, the suspended macrophytes foul beaches with their decaying mass. Changes in light penetration and the frequency of storms will have a great influence on water stratification. The potential for both sediment and pseudofeces resuspension, dislodging of macrophytes and movement of organic deposits onto the beaches will increase. In addition to the fouling of beaches, these events result in taste and odor problems in drinking water from geosmins and 2-methylisoborneol (Brownlee et al. 1984). Cleaning of beaches and treatment of water with activated carbon alleviates some local problems, but prevention of introduced species is the only management option to protect the integrity of the ecology of the Great Lakes.

Such problems are global and not restricted to freshwater. Another example is the introduction of the *Heterocapsa* species from tropical waters into Japan. This alga is now killing many of the oyster beds. The oysters sense the toxic algae, stop feeding and eventually die (Oda et al. 2000).

Efforts have been made to exchange ballast in the open sea, but it is not clear that such exchange is done well enough to remove the threat of introduced species. It is possible to pass such water thorough a hydrocyclone and then a membrane filter but the treatment costs would increase shipping expenses significantly. In the Great Lakes, proposals for transferring foreign cargo at marine ports to a fleet to service the Great Lakes are being evaluated. The international shipping fleets probably consider such proposals radical but so is the continued mongrelization of the ecology of the Great Lakes by their ballast water.

Nutrients

Some of the introductions of new species are complicated and not all changes occur via shipping. Birds are likely capable of moving algae between lakes. The movement of toxic algae around the world must also reflect global changes in conditions that enhance their growth. There are both direct effects such as nutrients and indirect changes such as sulfur loading or UV light.

Nitrogen enrichment is increasing globally. The transboundary fluxes of ammonia have resulted in its inclusion in the 1999 Gothenburg Protocol of the Un-ECE Convention of Long Range Transboundary Air Pollution (Sutton 2002). Nitrogen fertilization can increase nitrous oxide emissions (McCarl and Schneider 2001). Charlton et al. (1999) observed that nitrogen concentrations in Lake Erie have increased 2-6 fold since 1970. In Lake Superior, nitrogen concentrations have doubled and most of the input is likely atmospheric. The linkage between enhanced nitrogen supply and toxic algae is very strong. Blue-green algae like *Microcystis* cannot fix nitrogen and are clearly stimulated by enhanced nitrogen availability (Lean et al. 1982, Nalewajko and Murphy 2001). Production of the liver toxin microcystin by *Microcystis* has been shown to be positively correlated to nitrogen availability (Sivonen 1990).

Toxic algal blooms have been occurring more frequently throughout the world (Anderson et al. 1993, Hallegraeff 1993, Carmichael 1994). *Microcystis* production of microcystin is the most commonly documented problem. The recent deaths of 55 people in Brazil from microcystins may be the most alarming occurrence, but human health has been affected by algal toxins for some time (Falconer 1999). The toxin-producing alga *Cylindrospermopsis* is spreading north from tropical lakes (Chapman and Schelske 1997) and has reached Indiana (St. Amand 2002).

Other toxins such as thiaminase have been associated with *Microcystis* blooms (Hinterkopf et al. 1999). Thiaminase is an antimetabolite that can destroy the vitamin thiamin and in turn cause Haff disease (Buchholz et al. 2000). People can die from this disease, but most insight into this antimetabolite is associated with fisheries. Thiaminase activity can be high in introduced fish species. For example, alewives are rich in thiaminase and the introduction of alewives into the Great Lakes resulted in thiamin deficiency in salmonids and is now a rate-limiting factor on their production (Fitzsimons et al. 1999, Brown and Honeyfield 2000). Since beri beri disease in humans (thiamin deficiency) is present in some third world areas, the effect of thiaminase activity in lakes and oceans warrants more analysis. In North America, hatcheries have used thiamin treatment to enhance salmonid production but similar efforts are required with people too. The technical complexity is beyond the resources of some third world countries and global coordination of study and treatment is required.

The presence of more than one toxin or the ability of a toxin to enhance disease is not uncommon. Toxic algal blooms and avian botulism appear linked. Reports of dead birds or fish and blue-green algae are common (Yoo et al. 1995). Toxic blue-green algae in the prairies of North America are associated with outbreaks of avian botulism that have resulted in millions of dead birds (Yoo et al. 1995, Murphy et al. 2000). Unfortunately, botulism appears to be increasing. An outbreak of Type E Botulism in the lower Great Lakes began in 1998 and has continued in 1999, 2000, 2001 and 2002. Previous outbreaks were sporadic and the continuity of the current one is highly unusual. There are many indirect reasons why toxic algae are increasing too. Global increases in UV light are apparently stimulating microcystin production (Wulff 2001). This latter change has particular relevance to the Great Lakes, where water filtering by introduced species like zebra (*Dreissena polymorpha*) and quagga mussels (*Dreissena bugensis*) has dramatically increased light penetration. *Microcystis* species are enhanced in this environment for many reasons, including their ability to resist UV degradation. Microcystin is broken down by UV light (Tsuji et al. 1995) and one of its functions may be to protect the cell from UV damage.

Treatment of Toxic Algae

The global viewpoint on toxic algae has to be source control, but such actions to control nitrogen release or UV light would likely be byproducts of more basic changes required to prevent global warming, which is discussed well elsewhere (Nordhaus 2001, Walther et al. 2002). Induced flocculation of toxic algae might seem a local response but there is a global need to develop more cost-effective treatments. Removal of algae from lakewater has been done with lime or alum commercially for decades (Babin et al. 1989, Babin et al. 1994, Murphy et al. 1990, Prepas et al. 1990, and Serediak et al. 2002). The Japanese and Chinese also use various clays to remove toxic algae from marine water (Shirota 1989). Recently, the Australians have used a reactive clay that they call "Phoslock" to control algal blooms (Robb and Douglas 2000). The treatment process must protect the workers and ensure that algal cells are not lysed during the treatment. Some lake treatments, particularly copper, have resulted in lysis of cells and release of microcystin into solution (Jones and Orr 1994). The treatment dose must be no larger than needed to ensure rapid sedimentation and ideally restrict algal resuspenion during storms. Tracking the treatment in an open system is considerably more complex than treating all of a small lake. Typically in large lakes and the coastal ocean, plumes of toxic algae move with currents, and successful management will require an expert system or model to guide field crews. Satellite tracking might provide warning of blooms about to impact on either water intakes or aquaculture.

Eutrophication via Sulfur Loading

The enhancement of eutrophication by sulfur loading had been demonstrated in New England (Caraco et al. 1989), Europe (Bellemakers and Maessen 1998), Australia (Donnelly et al. 1997) and Japan (Murphy et al. 2001). This effect is strongest in softwater. Sites with high sulfate content like the ocean already recycle phosphorus faster than freshwater. Analysis of sediments from Lake Biwa, Japan indicates dissolution of phosphorus associated with increased sulfide and decreased redox potential in summer. The sediments might have lost some of their capacity to retain phosphorus. At one time, the high iron loading was able to stabilize vivianite (ferrous phosphate, $Fe_3(PO_4)_2 \cdot 8H_2O$) in the sediments. The following reaction represents how sulfur loading (with organic matter) results in the dissolution of phosphorus:

$$3SO_4^{-2} \leftrightarrows 3S^{-2} + Fe_3(PO_4)_2 \leftrightarrows 3FeS + 2PO_4^{-3}$$

The reactions are complicated by adsorption, especially to ferric iron. In poorly buffered softwater lakes, the pH increases when sulfate is reduced and phosphorus can desorb from ferric iron. The process can be greatly increased when the surface ferric iron is reduced.

Now that this sulfur-mediated reaction has been identified, the next management action should be the development of a sulfur-loading budget. With an estimate of historic and current loading, a treatment procedure can be quantified. Realistically not all sources can be controlled, but considerations of source control must part of the planning. It is not clear if the major Japanese sulfur sources are from agriculture, sewage, or combustion of fuels and wastes. Atmospheric loading of sulfur has likely increased in proportion to the increased amount of Japanese waste incineration and perhaps long-range transport of sulfur from sites in continental Asia. The potential for the involvement of similar mechanisms in enhanced eutrophication of other Asian lakes is high. The potential for continued or even more rapid release of phosphorus from the sediments of Lake Biwa is also high.

To control this sulfur mediated dissolution of vivianite would best be done through global efforts to reduce sources. Treatment of air pollution should be a primary objective of international agreements for many reasons other than minimizing this form of eutrophication. Some multinational firms resist change, challenge the science associated with climate change and at times threaten. When Environment Canada was pushing for reduced sulfur in gasoline, industry warned that gas supplies could be disrupted.

Direct treatment of sulfur-mediated eutrophication could be achieved by either oxidation of sulfide or complexation with iron. It has been demonstrated in lakes (Olsson et al. 1997, Hupfer et al. 1998) and groundwater (Griffioen 1994) that the formation of vivianite is partly controlled by the ratio of iron to phosphate and that as long as the iron concentration exceeds sulfide formation, vivianite will be stable. A most graphic illustration of porewater release from the sediments of Lake Biwa can be found in the results of an enclosure experiment where nitrate was injected into sediments to oxidize sulfides and precipitate phosphorus. Initially porewater phosphorus concentrations in sediments were similar in treated and control enclosures. Oxidation of sulphides in Lake Biwa sediments precipitated >80% of the porewater phosphorus (Murphy et al. 1999). The concentration of total phosphorus in the water column of the control enclosures increased to 45 $\mu g\ L^{-1}$ but remained less than 10 $\mu g\ L^{-1}$ in the treated enclosure.

Large-scale treatment of sulfide in sediments has been accomplished in Hong Kong using nitrate injection. In 1998 sediments were successively treated in situ (50 m by 50 m) (Babin et al. 1999). In 2001-2002, this same team treated 22 hectares of sediments in Hong Kong in situ with nitrate injection to control odour.

Abuse to Wetlands by Sulfur

In recent years, the eutrophication of wetlands and toxicity of sulfide to wetlands has been well documented. Dissolution of vivianite is certainly one common mechanism but sulfide itself is toxic and creates odour problems in and around wetlands too. In the Netherlands, naturally high levels of iron in groundwater provide some protection from formation of toxic sulfide and phosphorus dissolution in some wetlands but not others (Lamers et al. 2002). Sherwood and Qualls (2001) proposed pumping iron solutions into the storm sewer outlets in Florida to alleviate sulfur-mediated damage to wetlands. Iron solutions are available from primary steel mills at modest cost but care must be taken when using iron solutions from electric steel mills or others that use recycled metals. Their iron byproducts are often contaminated with other metals. Care would also have to taken to minimize effects on fisheries. Obviously there will be some negative aspects to any local treatment. In extreme plans, some wetlands may be dredged as a restoration tool. As soon as the organic content of the wetland returns, and as long as sulfur loading remains high, the wetland will return to its previous condition with odor problems, nutrient release, dead plants etc. The preferred action is source control, but there is a global need to develop better local remediation technologies.

Acid Rain

Acid rain corrodes buildings, impairs agriculture, damages forests, kills aquatic life and impairs human health. Usually most people associate sulfur pollution with acid rain. It is still a global problem (Satake et al. 2001). An integrated acid rain assessment model predicted that current sulfur emission controls would reduce damage in Ontario and Quebec but have little effect in Atlantic Canada (Jeffries et al. 2000). The recent resolution of sulfur-enhanced eutrophication is yet another reason to control sulfur discharge. Full implementation of current controls will still result in suboptimal pH values in many lakes in Ontario and Quebec. Some lakes have been restored by lime additions, but at times this produces aluminum toxicity. The aluminum toxicity may be transitory, but if the watershed is still impacted by acid rain, the aluminum problem may reoccur. Moreover, many lakes are difficult to access.

Global Remediation Technology Development/Regulation

Using scientific insight to enhance environmental management requires appropriate regulations. For example, we could utilize the insights into sulfur loading and the dissolution of vivianite. Currently, regulations attempt to prevent acidification but should be lower to prevent the sulfur loading exceeding a threshold where vivianite dissolution occurred. Each basin could have its own buffering potential and thus resilience to pollution. However, such detailed management is not likely to happen, especially in developing countries. For global contaminants such as sulfur, the regulations should be set to protect sensitive ecosystems.

Other contamination can be more complex than sulfur and international cooperation is required to improve management. The hazards of organic contaminants such as PAHs are difficult to assess. PAHs vary greatly in bioavailability. Bioremediation is best at removing the most bioavailable fraction of contaminants. The residuals may be almost refractory, but if their concentration exceeds a regulatory value, the treatment is judged to have failed. Since PAHs are carcinogens, the public fear is high. New methods of analysis, treatment and regulatory standards need to be linked.

The regulatory aspects of arsenic also require more international exchange. Recently the US EPA and the President of the USA disagreed on implementation of a new lower standard for arsenic in drinking water. Again public concerns over cancer drive much of the intensity of the exchange of opinions. Most of the risk analysis for setting the arsenic standard used Taiwanese data. More recent interpretation of this data suggests there may be a threshold before toxic responses are induced (Guo 2003). Furthermore data from western China has some interesting differences from the Taiwanese response (Wang and Wang 2003).

It is easy to find many examples where regulations in advanced countries are either not fully implemented or are flawed by interpretation. In Hamilton Harbour, Canada, steel mills are allowed to spray coal piles with waste oil (Curran et al. 2000). It suppresses dust. However, because this effluent is not viewed as a discharge, these steel mills are allowed to let the aqueous runoff leave their property untreated. The study of the steel mill runoff was an unusual example of industry and federal government cooperation but most regulatory control in Canada is done by provincial governments. In this case, the interpretation of pollution needs to be at a higher level, perhaps binational. Industries need to be regulated with consistent policies and be given simple tools to manage their problems.

Global Toxicants

Management of persistent organic pollutants is being addressed internationally. Countries like Germany and the USA banned production of PCBs. Although PCBs are no longer manufactured in most countries, they continue to recycle in nature. Whenever possible, sources must be controlled. Treatment of contaminated sites on land and in sediments must be considered. In water, dredging is the usual approach but in situ capping is also used. Isolation is always a consideration and new more effective types of in situ sediment capping are needed. Prevention is still better. Changes in incineration or product formulation may minimize dioxin formation in combustion (www.eco-logic-intl.com). Other toxins such as mercury are more complex and its mobility makes it truly a global contaminant (Schroeder and Munthe 1998). Mercury is naturally present, and people have some ability to excrete it. However, a diet that is natural but based upon too much fish can cause serious health impairment (Dickman et al. 1998). Moderation in eating certain fish is required. Such advice is not easy for poor communities that depend upon fish for protein. Moreover, in many places the enhanced atmospheric deposition of mercury is certainly not natural and neither is the mercury in the fish (Pirrone et al. 1998). One treatment that should be considered for hotspots is suppression of bacterial activity, which mediates methylation of mercury. A global perspective requires treatment of the source but implementation is often difficult. In places, natural gas is rich in mercury (Lewis 1995). It is removed to protect the gas refinery from metal corrosion, but at times it is put back into the distribution system. The alternative management would have to be the permanent fixation and storage of mercury, perhaps as an amalgam. Management of such wastes forever is an onerous obligation. Will multinationals continue to be responsible after the oil and gas has been depleted? Appropriate bonds would be required to ensure compliance and current pricing should provide for future management of wastes. It is a global problem and international agreements are required.

The international community must cooperate for global remediation. Weaker countries often do not have the resources required for complex measurement and analysis. Perhaps more important is the need to conform to market forces. It is a global economy and fears of being marginalized make resolution of issues such as climate change or mercury emissions difficult. The global interdependence of trade, investment, communication, transportation must extend to encompass global management of the environment. International cooperation in research should be coordinated with development of international regulatory policies.

References

ANDERSON, D. M., S. B. GALLOWAY, AND J. D. JOSEPH. 1993. Marine biotoxins and harmful algae: A national plan. WHOI Tech. Rep. 93-02, Woods Hole Oceanog. Inst., Woods Hole, MA.

BABIN, J. M., E. E. PREPAS, T. P. MURPHY, AND H. HAMILTON. 1989. A test of the effects of lime on algal biomass and total phosphorus concentrations in Edmonton stormwater retention lakes. Lake Reserv. Manage. **5**(1):129-135.

________, ________, ________, M. SEREDIAK, AND P. J. CURTIS. 1994. Impact of lime on sediment phosphorus release in hardwater lakes: the case of hypertrophic Halfmoon Lake, Alberta. The University of Alberta, Edmonton, Alberta. Lake Reservoir. Man. **8**:(2)1-15.

________, T. P. MURPHY, AND J. T. LYNN. 1999. *In situ* sediment treatment in Kai Tak Nullah to control odours and methane production, p. 823-828. *In* J. H. W. Lee, A. W. Jayawardena, and Z.Y. Wang [eds.], Proceedings Second International Symposium on River Sedimentation, Hong Kong, Dec. 16-18, A. A. Balkema Press.

BELLEMAKERS, M. J. S., AND M. MAESSEN. 1998. Effects of alkalinity and external sulfate and phosphorus load on water chemistry in enclosures in an eutrophic shallow lake. Water Air Soil Poll. **101:** 3-13.

BROWN, S., AND DALE HONEYFIELD. 2000. Report on Early Mortality Syndrome Workshop, Great Lakes Fishery Commission. Board of Technical Experts, Research Task Report, Ann Arbor, MI, USA.

BROWNLEE, B. G., D. S. PAINTER, AND R. J. BOONE. 1984. Identification of taste and odour compounds from western Lake Ontario. Water Poll. Res. J. Canada **19:** 111-118.

BUCHHOLZ, U., E. MOUZIN, R. DICKEY, R. MOOLENAAR, N. SASS, AND L. MASCOLA. 2000. Haff Disease: From the Baltic Sea to the U.S. Shore. 2000. http://www.cdc.gov/ncidod/eid/vol6no2/buchholtz.htm

CARACO, N. F., J. J. COLE, AND G. E. LIKENS. 1989. Evidence for sulfate-controlled phosphorus release from sediments of aquatic systems. Nature **341:** 316-318.

CARMICHAEL, W. 1994. The toxins of cyanobacteria. Scientific American **270**(1): 78-84.

CHAO, B. F. 1995. Anthropogenic impact on global geodynamics due to reservoir water impoundment. Geophys. Research. Letter. **22:** 3529-3532.

CHAPMAN, A. D. AND C. L. SCHELSKE. 1997. Recent appearance of *Cylindrospermopsis* (Cyanobacteria) in five hypertrophic Florida lakes. J. Phycology **33:** 219-226.

CHARLTON, M. N., R. LESAGE, AND J. E. MILNE. 1999. Lake Erie in Transition: the 1990's, p. 97-123. *In* M. Munawar, T. Edsall and I. F. Munawar [eds.], In State of Lake Erie (SOLE) – Past, Present and Future. Bachhuys Publishers. Leiden, the Netherlands.

CURRAN, K. J., K. N. IRVINE, I. G. DROPPO, AND T. P. MURPHY. 2000. Suspended solids, trace metal and PAH concentrations and loadings from coal pile runoff to Hamilton Harbour, Ontario. J. Great Lakes Res. **26**(1):18-30.

DICKMAN, M. D., C. K. M. LEUNG, AND M. K. H. LEONG. 1998. Hong Kong male subfertility links to mercury in human hair and fish. Sci. Total Environ. **214:** 165-174.

DONNELLY, T. H., M. R. GRACE, AND B. T. HART. 1997. Algal blooms in the Darling-Barwon River, Australia. Water Air Soil Pollut. **99:** 487-496.

FALCONER, I. R. 1999. An overview of problems caused by toxic blue-green algae (cyanobacteria) in drinking and recreational water. Environ. Toxicol. **14:** 5-12.

FITZSIMONS, J., S. B. BROWN, D. C. HONEYFIELD, AND J. G. HNATH. 1999. A review of early mortality syndrome (EMS) in Great Lakes salmonids: relationship with thiamin deficiency. Ambio. **28**(1): 9-15.

GRIFFIOEN, J. 1994. Uptake of phosphate by iron hydroxides during seepage in relation to development of groundwater composition in coastal areas. Environ. Sci. Technol. **28:** 675-681.

GUO, H-R. 2003, Arsenic in Drinking Water and Cancers in Taiwan. *In* T. Murphy and J. Guo [eds.], Aquatic Arsenic Toxicity and Treatment, Backhuys Press, in press.

HALLEGRAEFF, G. M. 1993. A review of harmful algal blooms and their apparent global increase. Phycologia **32**(2): 79-99.

HINTERKOPF, J. P., D. C. HONEYFIELD, J. MARKAREWICZ AND T. LEWIS. 1999. Abstract. Abundance of Plankton Species in Lake Michigan and Incidence of Early Mortality Syndrome from 1983 to 1992. Report on Early Mortality Syndrome Workshop, Great Lakes Fishery Commission. Board of Technical Experts, Research Task Report, Ann Arbor, MI, USA.

HUPFER, H., P. FISCHER, AND K. FRIESE. 1998. Phosphorus retention mechanisms in the sediment of an eutrophic mining lake. Water Air Soil Poll. **108:** 341-352.

JEFFRIES, D. S., D. C. L. LAM, I. WONG, AND M. D. MORAN. 2000. Assessment of changes in lake pH in southeastern Canada arising from present levels and expected reductions in acidic deposition. Can. J. Fish. Aquat. Sci. **57**, Suppl. 40-49.

JONES, G. J. AND P. T. ORR. 1994. Release and degradation of microcystin following algicide treatment of a *Microcystis aeruginosa* bloom in a recreational lake, as determined by HPLC and protein phosphatase inhibition assay. Wat. Res. **28**(4): 871-876.

KARIM, M. M. 2000. Arsenic in groundwater and health problems in Bangladesh. Water Resources **34**(1): 304-310.

LAMERS, L. P. M., S. J. FALLA, E. M. SAMBORSKA, I. A. R. VAN DULKEN, G. VAN HENGSTUM, AND J. G. M. ROELOFS. 2002. Factors controlling the extent of eutrophication and toxicity in sulfate-polluted freshwater wetlands. Limnol. Oceanogr. **47:** 585-593.

LEAN, D. R. S., T. P. MURPHY, AND F. R. PICK. 1982. Photosynthetic response of lake plankton to combined nitrogen enrichment. J. Phycol. **18:** 509-521.

LEWIS, L. 1995. Measurement of mercury in natural gas streams. Proceedings of the Seventy-Fourth Gas Processors Association Convention, San Antonio, Texas.

MCCARL, B. A. AND U. A. SCHNEIDER. 2001. Greenhouse gas mitigation in U.S. agriculture and forestry. Science **294:** 2481-2482.

MURPHY, T. P., E. E. PREPAS, J. T. LIM, J. M. CROSBY, AND D. T. WALTY. 1990. Evaluation of calcium carbonate and calcium hydroxide treatments of prairie drinking water dugouts. Lake Reservoir. Man. **6**(1):101-108.

________, A. LAWSON, M. KUMAGAI, AND J. BABIN. 1999. Review of Canadian Experiences in Sediment Treatment. Aquatic Ecosystem Health & Management **2:** 419-434.

________, __________, C. NALEWAJKO, H. MURKIN, L. ROSS, K. OGUMA, AND T. MCINTYRE. 2000. Algal toxins – initiators of avian botulism? Environ. Tox. **15:** 558-567.

________, __________, __________, AND M. KUMAGAI. 2001. Sediment Phosphorus Release in Lake Biwa, Limnology **2:** 119-128.

NALEWAJKO, C., AND T. P. MURPHY. 2001. The importance of temperature, and N and P availability to the abundance of *Anabaena* and *Microcystis* in Lake Biwa, Japan: an experimental approach. Limnology **2:** 45-48.

NIKIFORUK, A. 2002. The Water Crisis, Alberta Venture Magazine.

NORDHAUS, W. D. 2001. Global warming economics. Science **294:** 1283-1284.

ODA, T., Y. SATO, D. KIM, T. MURAMATSU, Y. MATSUYAMA, AND T. HONJO. 2000. Hemolytic activity of Heterocapsa circularisquama (Dinophyceae) and its possible involvement in shellfish toxicity. J. Phycol. **37:** 509-516.

OLSSON, S., J. REGNÉLL, A. PERSSON, AND P. SANDGREN. 1997. Sediment-chemistry response to land-use change and pollutant loading in a hypertrophic lake, southern Sweden. J. Paleolim. **17:** 275-294.

PIRRONE, N., I. ALLEGRINI, G. J. KEELER, J. O. NRIAGU, R. ROSSMAN, AND J. A. ROBBINS. 1998. Historical atmospheric mercury emissions and deposition in North America compared to mercury accumulations in sedimentary records. Atmospheric Environment **32:** 929-940.

PREPAS, E. E., T. P. MURPHY, J. M. CROSBY, D. T. WALTY, J. T. LIM, J. BABIN, AND P. A. CHAMBERS. 1990. Reduction of phosphorus and chlorophyll *a* concentrations following $CaCO_3$ and $Ca(OH)_2$ additions to hypertrophic Figure Eight Lake, Alberta. Environ. Sci. Tech. **24:** 1252-1258.

REID, J. 1994. Arsenic occurrence: USEPA seeks clearer picture. J. AWWA **86**(9): 44-51.

RICCIARDI, A. 2001. Facilitative interactions among aquatic invaders: is an "invasional meltdown" occurring in the Great Lakes. Canadian Journal of Fisheries and Aquatic Sciences **58**(12): 2513-2525.

ROBB, M., AND G. DOUGLAS. 2000. Clay treatment – Australia tests "Phoslock" river treatment. Scope Newsletter 38 and http://www.csiro.au and type "Phoslock" in search.

SATAKE, K., S. KOJIMA, T. TAKAMATSU, J. SHINDO, T. NAKANO, S. AOKI, T. FUKUYAMA, S. HATAKEYAMA, K. IKUTA, M. KAWASHIMA, Y. KOHNO, K MURANO, T. OKITA, H. TAODA, AND K. TSUNODA. 2001. Acid Rain 2000 - Conference Summary Statement - Looking Back to the Past and Thinking of the Future. Water Air Soil Pollut. **130:** 1-16.

SCHROEDER, W. H., AND J. MUNTHE. 1998. Atmospheric mercury – an overview. Atmospheric Environment **32:** 809-822.

SCOTT, M. C. AND G. S. HELFMAN. 2002. Native invasions, homogenization, and the mismeasure of integrity of fish assemblages. Fisheries **26**(11): 6-15.

SEREDIAK, M., E. E. PREPAS, T. P. MURPHY, AND J. BABIN. 2002. Development, construction and use of lime and alum application systems in Alberta. Lake Reserv. Manage.

SHERWOOD, L. J. AND R. G. QUALLS. 2001. Stability of phosphorus within a wetland soil following ferric chloride treatment to control eutrophication. Environ. Sci. Technology **35:** 4126-4131.

SHIROTA, A. 1989. Red tide problem and countermeasures. Int. J. Aquat. Fish. Technol. **1:** 195-223.

SIVONEN, K. 1990. Effects of light, temperature, nitrate, orthophosphate and bacteria on growth of and hepatotoxin production by *Oscillatoria agardii* strains. Applied Environ. Microbiol. **56:** 2658-2666.

SMEDLEY, P. L. AND D. G. KINNIBURGH. 2002. A review of the source, behaviour and distribution of arsenic in natural waters. Applied Geochemistry **17:** 517-568.

ST. AMAND, A. 2002. *Cylindrospermopsis*: an invasive toxic alga. Lakeline **22**(1): 36-37.

SUTTON, M. A. 2002. Introduction: fluxes and impacts of atmospheric ammonia on national, landscape and farm scales. Environmental Pollution **119:** 7-8.

TSUJI, K., T. WATANUKI, F. KONDO, M. F. WATANABE, S. SUZUKI, H. NAKAZAWA, M. SUZUKI, H. UCHIDA, AND K. HARADA. 1995. Stability of microcystins from cyanobacteria - II. Effect of UV light on decomposition and isomerization. Toxicon. **33**(12): 1619-1631.

WALTHER, G. R., E. POST, P. CONVEY, A. MENZEL, C. PARMESAN, T. J. C. BEEBEE, J. M. FROMENTIN, O. HOEGH-GULDBERG, AND F. BAIRLEIN. 2002. Ecological responses to recent climate change. Nature **416:** 389-395.

WANG, L., AND S. WANG. 2003. Arsenic in Water Its Health Effect. *In* T. Murphy and J. Guo, Aquatic Arsenic Toxicity and Treatment, Backhuys Press, in press.

WULFF, A. 2001. Is there a coupling between increased UVB radiation and toxic algal blooms. SIL 2001 XXVIII Congress, Melbourne, Australia, Abstract.

YOO, R. S., W. W. CARMICHAEL, R. C. HOEHN, AND S. E. HRUDEY. 1995. Cyanobacterial (Blue-green algal) Toxins: A Resource Guide. American Water Works Association 230 p.

3-2. Removing Environmental Contaminants with Aquatic Plants and Algae

Hans G. Peterson

WateResearch Corp., Box 7, Site 508, RR #5, Saskatoon, SK S7K 3J8, Canada

Abstract

Contamination of aquatic systems with nutrients, organic and inorganic compounds results in water quality degradation both in rivers and lakes. While natural systems have some capacity to self-purify by the activities of aquatic plants and algae there are many situations where constructed or engineered systems integrating photosynthetic organisms can increase the biological decontamination of different types of wastewaters by one or several orders of magnitude. Constructed aquatic plant treatment systems can be integrated into urban landscapes where contaminants in stormwater runoff are removed before the water is discharged to natural ecosystems. Similarly, more concentrated waste streams, such as sewage or manure streams can be processed by photosynthetic organisms. Contaminants from natural resource related activities, such as mining and the extraction of oil, can also be treated in specially designed water treatment systems based on aquatic plants. The contaminants can be removed from the water through the normal growth of the algae and aquatic plants including the uptake of the nutrient phosphorus and organic compounds. Non-biodegradable compounds, such as metals, can be concentrated by algae and aquatic plants and the purified water can then be returned to natural lakes and rivers or be safely used for other purposes. The advantages of contaminant removal with algae and plants include good removals, less use of chemicals, less cost, and the methods are environmentally friendly. There are also man-made reservoirs (ponds) for consumptive purposes and increasingly stringent regulations on the quality of drinking water makes it essential to manage such reservoirs in order to limit the concentration of deleterious substances. A common theme for treatment ponds is the requirement to use or manage aquatic plants and algae to further the specific goals of a particular treatment pond.

Introduction

The discussion in this chapter is limited to how wetland plants and algae can be used to clean up environmental contaminants either alone or in combination with more conventional treatment technologies. Issues that need to be considered and challenges that need to be resolved are aimed to highlight both the diversity of the problems and the complexity of some solutions. It must, however, be recognized that as we use biological systems to remove contaminants, the complexity of the solutions increase. For example, removal of phosphorus from a waste stream can be achieved using coagulation with alum or ferric chloride/sulphate. The chemical flocculation process can be modelled and tested in a couple of simple experiments. While this will allow us to discharge water into natural systems with low levels of P, we have at the same time generated a new environmental problem through the formation of alum or ferric sludges.

There are now microbial processes that are used in sewage treatment plants that can remove both P and N. However, even when the engineered sewage treatment systems are using microbes to accomplish specific goals, such as N, P, and BOD removals, there are large inputs of energy (mixing) or large inputs of chemicals (for example ethanol for nitrate removal by anaerobic bacteria). While these systems are steps in the right direction, more economical systems to deal with aquatic pollutants should be sought. What has also been largely forgotten in the treatment of sewage is the survival of "true" rather than "indicator" microbes. Human pathogens depend on their host for reproduction and will find wetlands very hostile environments with natural die-off, temperature, ultraviolet light, unfavourable water chemistry, predation and sedimentation all playing a part in pathogen removals (Kadlec and Knight 1996).

The use of photosynthetic, heterotrophic or photo-heterotropic algae and aquatic plants to degrade or remove contaminants or to produce biomass if carefully designed can provide environmentally friendly solutions in a cost-effective fashion. Combining such growth with other biological strategies, such as harvesting of algae (and pathogenic microbes) using zooplankton can bring further benefits to this. There are tremendous opportunities to explore the use of such organisms for the treatment of domestic sewage, animal manure and contaminants from different industries, stormwater runoff and indeed drinking water reservoirs.

Results and Discussion

Water treatment goals

Biological treatment processes depend on the ability of organisms to degrade organic material, remove nutrients, and toxic elements including heavy metals. The driving force to change the properties of a waste stream from one with high levels of organics, reduced N compounds, dissolved P, and heavy metals to one with low levels of these compounds generally rests with government regulations. This forces municipal treatment plants, industrial processing plants, and other waste generating facilities to comply with local/national discharge standards/guidelines. Regulators may also want to know about the potential toxicity of the effluents and their content of pathogenic microbes. The ultimate goal is to have an effluent that has a low impact on the receiving water.

Some industries produce wastewater streams that are highly elevated with specific contaminants and there are even limits for the concentration of compounds that can be discharged into municipal sewage treatment plants. If these limits are exceeded, surcharges will typically be levied. There are therefore many requirements to pre-treat wastewater streams even when they are discharged into sewage treatment plants.

The parameters of concern and the removal rates required shown in Table 1 were assembled by Peterson (1998) from the scientific literature. Conventional water treatment typically employs the addition of various chemicals, such as coagulants (aluminum and ferric salts for example),which remove some contaminants such as suspended solids, phosphorus and some dissolved organic compounds. This treated water is then settled or filtered before it is discharged into natural aquatic environments. Large quantities of sludge are produced during these processes and the organic material is bound to metal salts rather than oxidized and removed.

During biological treatment of waste material naturally occurring physical and chemical processes including adsorption, settling, precipitation and filtration are at play, but in addition the microbes, algae and higher plants are actively processing contaminants into other products. For example, during biological treatment ammonium can be converted to nitrate and particulate and dissolved organic material are broken

Table 1 Parameters of concern and requiring treatment in organic wastewaters.

Parameter	Common level in waste	Desired effluent level	Desired removal rates
Total suspended solids	150 mg L^{-1}	<30 mg L^{-1}	>80%
BOD5	150 mg L^{-1}	<30 mg L^{-1}	>80%
Ammonia-N	25 mg L^{-1}	<2.5 mg L^{-1}	>90%
Phosphorus	10 mg L^{-1}	<1 mg L^{-1}	>80%
Pathogenic microbes	10^4 L^{-1}		>99%
Indicator organisms (faecal coliforms)	10^9 L^{-1}		>99%
Odour and colour		Low levels	>90%
Toxic substances		Prohibited	>90%

down to carbon dioxide and water. Biologically active systems are also producing highly unfavourable environments for pathogenic microbes and their removal can proceed faster than in biologically inactive systems (Kadlec and Knight 1996).

This paper is addressing the use of aquatic plants to purify contaminated environments. The inclusion of biological techniques to treat water will require that the engineering profession invest more time in "environmental engineering". Maybe bio-engineering will be a more appropriate word for dealing with the construction of units specifically designed to treat waste products generated by man. Here the discussion is centered on how such "bio-engineered" systems may become treatment systems of the future.

Treatment with aquatic plants

Vascular aquatic plants can form wetlands that are among the most productive in the world. Combined with this high productivity are conditions that are generated resulting in the colonization by other organisms including algae, bacteria and protozoan organisms. All of these organisms play integral roles in the functioning of water treatment wetlands where conditions for organic matter degradation and the removal of metals and nutrients need to be optimized. These attributes are exploited in their use in constructed wetlands to treat different types of waste (Brix and Schierup 1989).

The fate processes in constructed wetlands were discussed by Tchobanoglous

Table 2 Fate processes in construction wetlands (Adapted from Tchobanoglous [1993]).

Processes	Comments
Bacterial conversion	Bacterial conversion (both aerobic and anaerobic) is the most important process that transforms contaminants discharged to constructed wetlands. The exertion of biochemical oxygen demand (BOD) is the most common example of bacterial conversion encountered in water quality management. The depletion of oxygen by the aerobic conversion of organic wastes is also known as deoxygenation. Solids discharged with treated wastewater are partly organic. Upon settling to the bottom they decompose bacterially, either anaerobically or aerobically, depending on conditions. The bacterial transformation of toxic organic compounds is also of great significance.
Gas absorption / desorption	The process whereby a gas is taken up by a liquid is known as absorption. For example when the dissolved oxygen concentration in a body of water with a free surface is below the saturation concentration in the water, a net transfer of oxygen occurs from the atmosphere to the water. This transfer (mass per unit time per unit surface area) is proportional to the amount by which the dissolved oxygen is below saturation. The addition of oxygen to water is also know as reaeration. Desorption occurs when the concentration of the gas in the liquid exceeds the saturation value and there is a transfer from the liquid to the atmosphere.
Sedimentation	The suspended solids discharged with treated wastewater ultimately settle to the bottom. Settling is enhanced by flocculation and hindered by ambient turbulence. In some wetlands, turbulence is often sufficient to distribute the suspended solids over the entire water depth.
Natural decay	In nature, contaminants will decay for a variety of reasons, including mortality in the case of bacteria and photo-oxidation for certain organic constituents. Natural decay generally follows first-order kinetics.

(1993, Table 2). The use of wetlands to treat municipal wastewater for small communities has several advantages. The cost of constructing treatment wetlands is typically 10 to 50% of the cost of conventional treatment (Brix and Schierup 1989). This has led the European Union to advocate the use of such wetlands for sewage treatment for communities that are smaller than 1,000 people (Cooper 1993). The U.S. Environmental Protection Agency has also encouraged the innovative use of wetlands (Bastian et al. 1989) although in North America such use is not widespread.

Table 2 Continued.

Adsorption	Many chemical constituents tend to attach or sorb onto solids. The implication for wastewater discharges is that a substantial fraction of some toxic chemicals is associated with the suspended solids in the effluent. Adsorption to solids followed by settling results in the removal from the water column of constituents which might not otherwise decay.
Volatilization	Volatilization is the process whereby liquids and solids vaporize and escape to the atmosphere. Organic compounds that readily volatilize are known as volatile organic compounds (VOCs). The physics of this phenomenon is very similar to gas absorption, except that the net flux is out of the water surface.
Chemical reactions	Important chemical reactions that occur in wetlands include hydrolysis, photochemical, and oxidation-reduction reactions. Hydrolysis reactions occur between contaminants and water. Solar radiation is known to trigger a number of chemical reactions. Radiation in the near-UV and visible range is known to cause the breakdown of a variety of organic compounds.

Wetland types and treatment efficiency

There are two major types of wetlands, surface flow and subsurface flow. Most natural wetlands are surface flow. Surface flow systems have an open water surface as part of the wetland and the risk of clogging is low. Treatment efficiency in both natural and constructed surface flow wetlands is often compromised due to lower substrate and pollutant interactions. Ammonia and phosphorus can, however, be effectively treated in such wetlands as reviewed by Kadlec (1995). Due to lower costs a larger-size surface flow wetland can be constructed and for different types of removal such wetlands may be the best option. Kadlec and Knight (1996) evaluated these issues and give helpful information for the construction of different types of wetlands.

Subsurface flow forces the water to take a tortuous path through a particulate medium, such as gravel, stones or sand, which when interspersed with roots from vascular aquatic plants generate effective filtration in addition to the reactions outlined in Table 2. The substrate size is generally around 0.6 cm to 2.5 cm with hydraulic conductivities of 10,000-50,000 m/d (Reed et al. 1995). In subsurface flow wetlands the interaction between substrate and pollutant is vastly increased, but as their effectiveness increase so do their tendency to plug. The surface of a subsurface flow wetland is typically covered by aquatic vegetation but the actual surface is dry most of the time. Subsurface flow wetlands are especially well suited to treat toxic and odorous compounds (Reed et al. 1995). The surface flow wetland (including natural wetlands) are better for wildlife habitat and cost, while the sub-surface flow wetland is better for the prevention of mosquitoes, odour, and human contact (Kadlec 1995).

A pollutant entering a wetland must leave by one or more routes, be chemically converted, or be stored in the wetland (Kadlec 1995). Treatment efficiency will depend on a host of factors including inlet concentration, water depth, dissolved oxygen concentration, temperature, loading rate, water flow, type of plants, and type of substrate etc. A continuous external input, such as animal or human waste containing nutrients and organic carbon is better able to maintain sustained high removal efficiencies (Whigham 1995). Due to the interactions of many different factors removal of BOD, nutrients and microbial contaminants can be highly variable ranging from very low (around 20%) to very high (>80%). The two most studied pollutants in wetlands are nitrogen and phosphorus and the retention of these compounds by different types of wetlands was compiled by Mitsch and Gosselink (1993, Figure 1).

Natural wetlands have not been tested at the high loading rates often introduced into constructed wetlands. Inputs of wastewater to natural wetlands will probably result in loss of biodiversity and downstream water quality can also be negatively affected. It is only in rare circumstances that natural wetlands are suitable for wastewater treatment use and then only if the waste has been extensively pre-treated (Ryding and Rast 1989). There is a need to pre-treat wastewater before introducing it into any type of wetland especially removing particles to avoid clogging of the bed.

The size of treatment wetlands can range from very small to very large (0.02 to 500 ha, Knight et al. 1993) covering treatment requirements for single family dwellings to treatment requirements of towns and cities. Plans for large-scale P removal wetlands include reducing the P discharged into the Florida Everglades by 80% highlighting the future potential of water treatment using aquatic vascular plants (Stone and Legg 1992).

Aquatic plant species

The issues surrounding the expansion of wetlands as strategies for widespread pollution control rests with our ability to use vascular aquatic plants as part of more strategies to remove contaminants from the environment. First, constructed wetlands typically employ a small number of vascular aquatic plants. In Europe for waste treatment wetlands the common reed (*Phragmites* sp.) is the preferred species while in North America cattails (*Typha* sp.) are most frequently used.

For the subsurface flow wetlands, which are constructed with coarse material to allow for water flow through the substrate, the plants especially in northern climates

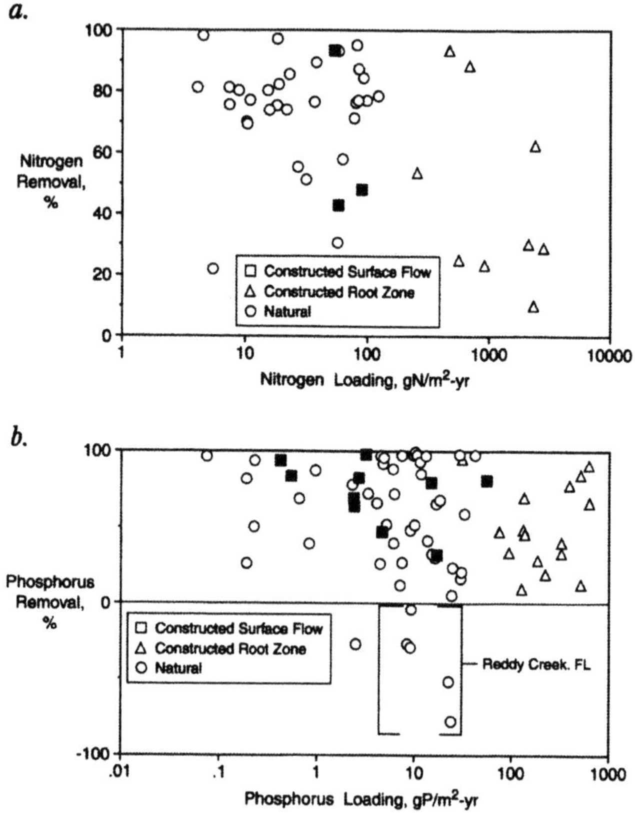

Fig. 1. Nitrogen (a) and phosphorus (b) removal by different types of wetlands receiving wastewater or river water as a function of loading (adapted from Mitsch and Gosselink [1993]).

face harsher conditions than what the same plants would encounter in natural wetlands. To test whether different types of plants would survive in Canada, a pilot-scale wetland (20 by 40 m) was constructed by the City of Regina in 1992. This wetland was designed to remove algae and phosphorus from a stormwater pond (runoff from city streets and properties).

Plant survival is shown in Table 3. Plantings carried out in the fall were much less successful than early summer plantings. A planting density of 4% was sufficient. "Mud-balls" (20 cm cubes of plant material) were obtained from local wetlands and were planted 5-10 cm below the top of the gravel. Once the hole was dug, a peat moss container with slow release fertilizer was put into the hole, then the mud ball was put in

Table 3 Survival of plants in a hostile environment.

Macrophytes		Survival	
Species	**% Survival 1993 Fall 1992 planting**	**% Survival 1994 Early summer 1993 planting**	**% Survival 1995 Early summer 1994 planting**
***Sagittaria* sp. (arrowhead)**	18	85	>90
***Eleocharis* sp. (spikerush)**	0	100	>90
***Carex* sp. (tall grass)**	0	7 (was planted in the fall)	~70
***Scirpus* sp. (bullrush)**	50	69	~70
***Typha* sp. (cattails)**	16	100	>90
***Phragmites* sp. (common reed)**	48	47	>90
***Salix* sp. (willow)**	5	--	--

and covered with gravel. Within two years the coverage averaged 23%, and it then proceeded rapidly to more than 90% coverage. The potential for expanding the number of species in wetland treatment processes is therefore considerable.

This has been used by the City of Regina during the 2002 field season where spikerush (*Eleocharis* sp.), bullrush (*Scirpus* sp.), and the common reed (*Phragmites* sp.) have been planted along the shores of 4 stormwater ponds. For plantings into existing shores around the stormwater ponds we have not added any fertilizer to the mud ball. Excellent growth has been seen and these plants will become a "passive treatment system". The shorelines of stormwater ponds are frequently lined with stones (as is the case in Regina). There is also a legal requirement to construct a shallow fringe around the ponds to prevent accidents where children fall into the ponds and immediately get out of depth. Around 5 m from the shore the depth increases more rapidly.

This 5 m zone around the ponds is highly visible and frequently the location for massive developments of nuisance algae, such as macroalgae (*Enteromorpha* etc.). By planting selected aquatic plants along the shore algae attachment can be discouraged and the wave action will bring nutrients to the shore, penetrate the stones and reach the roots of these near-shore planted macrophytes. In addition, shore-line planted macrophytes will prevent wind and wave-induced shoreline erosion and its concomitant input of nutrients to the open water system. During 2003 each pond in Regina will be planted with aquatic macrophytes along the shoreline (only around 1% coverage) in an attempt to reach more than 50% coverage within the following three years.

The above illustrates the scope for expanding the use of different species of aquatic macrophytes to remove aquatic pollutants. It also illustrates how the use of such macrophytes can be expanded. It is also clear that not all algae and aquatic plant growth in urban lakes is desirable and there is a need to manage both pollutants and aquatic plant populations. Further issues that need to be expanded on include year-round operation in cold climates, effective removal of pathogens, and incorporating features that will enhance pollutant removals (e.g., substrates that can adsorb pollutants). Other issues include combining different waste streams that may enhance each other for removal of different pollutants, environmental modifications, such as aeration of wetland cells etc. The wetland treatments may therefore become more complex or as described above, less complex, when it comes to plantings of macrophytes along shore lines of polluted water bodies.

Treatment ponds

Treatment ponds are defined as man-made open water bodies including drinking water reservoirs, sewage or industrial waste lagoons, and stormwater ponds draining urban landscapes. The fringes of some of these ponds can be planted with aquatic plants as described above to make use of the stabilizing properties and nutrient uptake capabilities of aquatic plants. In treatment ponds, however, lake management strategies become increasingly important with increasing open water size. The treatment ponds will, for most part, be discharged into a natural river or lake and the goal is to improve the quality of the water in the pond so that it can be discharged without effects on the natural system. For drinking water reservoirs the objective is to improve the quality with respect to compounds affecting consumptive uses. As these treatment ponds are man-made and are not considered "natural" our ability to manage them is far greater than natural lakes.

The strategies used include those of regular lake management, such as biomanipulation where strategies to control different organisms while others are promoted are used (this is extensively covered in this book in chapter 5-1). Traditionally various chemicals, such as copper sulphate and pesticides, such as diquat have played prominent roles in treatment pond management. Here, examples of the issues confronting managers of treatment ponds are highlighted with a case scenario from a drinking water reservoir.

In water short areas of the world the construction of reservoirs to conserve or dam water is an ongoing activity. These reservoirs serve as water for various purposes including consumption by humans, livestock as well as serving various industrial and agricultural processing needs. For example, in Western Canada there are more than 100,000 constructed drinking water reservoirs ranging from individual household size (a few million litres) to municipal reservoirs (tens of millions of litres). If these ponds are not managed, they will typically generate blooms of blue-green algae (cyanobacteria) as nitrogen is in short supply leaving the nitrogen-fixing blue-green algae with a competitive advantage. The goal with the drinking water reservoirs is to trap water and to preferably maintain or improve its quality during storage, but the shallow depth of most of these ponds (around 1.5 m average depth) leaves ample opportunities for aquatic plants and algae to proliferate. These ponds typically also drain agricultural land rich in nutrients and with soil erosion from ploughed fields in particular being a problem with the reservoirs receiving large quantities of particulate and dissolved nutrients.

We need, however, to deal with these ponds as treatment ponds forming the first step in the water treatment process. Interventions in these types of ponds need to minimize the growth of blue-green algae while the growth of other algae, such as green algae, are beneficial under some circumstances, but not in others. This will be illustrated by actual data from a drinking water pond, which was treated according to government instructions for how to deal with blue-green algae blooms. This example serves to illustrate how easily we are able to manage or mismanage treatment ponds.

Two pesticides are allowed for use in drinking water supplies in Canada, copper sulphate and diquat (common brand name in Canada, Reglone A). Copper sulphate's main use is for control of blue-green algae. Diquat is mainly used for weed control. Copper sulphate treatments of up to 1 kg/million L were recommended by government agencies across the Canadian prairies until information that is described below was documented by Hans Peterson and co-workers in various government reports during the late 80ies. If this quantity of copper sulphate is used then the maximum dissolved copper concentration in the treated water after thorough mixing would be 0.25 mg Cu/L. There are no restrictions on consumption of copper treated water as the drinking water guideline for copper is 1 mg Cu/L in most jurisdictions around the world. However, since copper treatments are designed to kill blue-green algae, which may release toxins, a delay in consumption is called for. For an individual user that delay may be appropriately set to 14 days. A user can comply with this by collecting enough

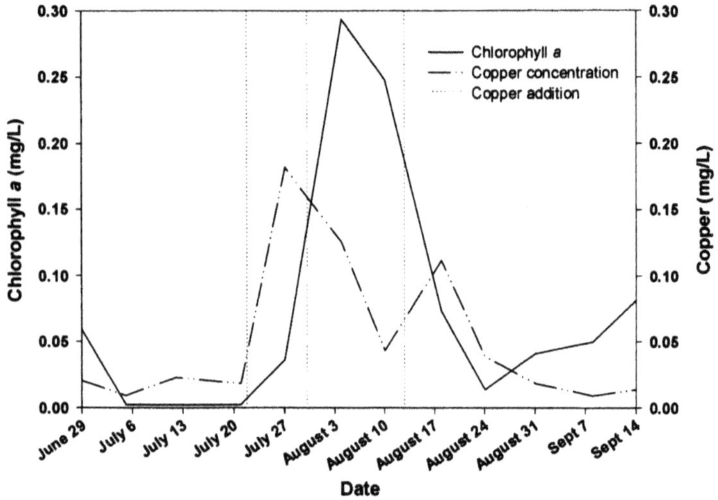

Fig. 2. Algal biomass (chlorophyll a) and dissolved copper in a Canadian drinking water reservoir from the end of June to mid September.

drinking water for 14 days before the pond is treated.

Copper sulphate treatments of a bloom of the nitrogen-fixing blue-green alga *Aphanizomenon* will be examined in some detail. This blue-green alga forms "finger-nail clippings" composed of large numbers of straight multi-cellular filaments. Even when the level of this alga is ten times lower than an "algal bloom" (defined as more than 50 micrograms chlorophyll a/L) it can be easily seen by the naked eye. Indeed, the owner of the drinking water reservoir noticed the "finger-nail clippings" and immediately applied a high dose of copper sulphate (1.1 kg/million L) and the *Aphanizomenon* was eradicated, but ten days after copper treatment the phytoplankton biomass had increased 50 times. This set of events is shown in Figure 2.

The dissolved copper levels in the drinking water reservoir drastically increased following the first application, but additional applications did not increase these levels further. The algae took up a large amount of the copper. The expected dissolved level of a chemical that is added to a water body will therefore strongly depend on the presence of algae and if these algae will interact with the compound. Small *Chlorella*-type green algae replaced the *Aphanizomenon* in the presence of copper. These algae were less than 5 micrometer in size and presented new challenges as these algae passed through the domestic filtration system. The tap water was lime-green and odorous.

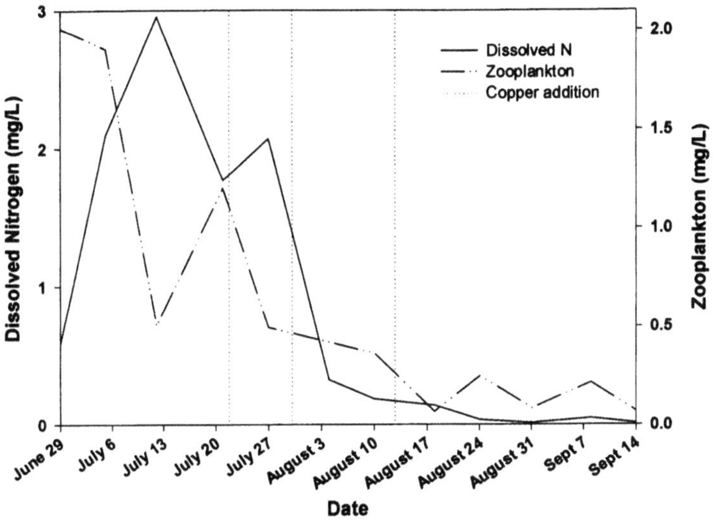

Fig. 3. Dissolved inorganic nitrogen and zooplankton in a Canadian drinking water reservoir from the end of June to mid September.

Two more applications of copper was made and after the third treatment the biomass of the algae finally decreased. Upon examining nutrient levels it became clear that phosphorus was at non-limiting levels, but the inorganic nitrogen was reduced to low levels when the bloom declined (Figure 3). Lack of readily available N was the most likely cause of the collapse of this bloom. The quantity of phytoplankton-grazing zooplankton was relatively high before the copper treatment even during the dominance of *Aphanizomenon*, which is almost inedible (Figure 3). The green algae that developed after copper treatment would have been an ideal source of food for the zooplankton. Instead, the zooplankton decreased to low levels despite the abundance of high quality food.

The management of this drinking water reservoir can be summed up as a successful destruction of the blue-green algae and the zooplankton. The accidental destruction of the zooplankton caused the proliferation of copper-tolerant green algae. This was not a desired outcome for a drinking water reservoir as several drinking water quality parameters deteriorated including turbidity, taste and odour, manganese and pH as well as wide fluctuations in ammonium and oxygen levels.

Indeed, the response of blue-green algae to chemicals commonly used in water treatment processes and surface water management can be classified into three types of compounds (Peterson et al. 1995). Effects were determined by the response of the nitrogen-fixing blue-green alga *Aphanizomenon flos-aquae* after exposure to the different chemicals. Type 1 compounds, aluminum sulphate, and ferric chloride, caused no physiological toxicity, cell membrane damage or the release of DOC at treatment usage concentrations. Type 2 and 3 chemicals caused physiological toxicity at or below concentrations used in water treatment operations. Type 2 chemicals, calcium hydroxide and hydrogen peroxide, caused limited membrane damage and DOC release. Type 3 chemicals, chlorine, copper sulphate, and potassium permanganate, caused extensive damage including the release of large quantities of DOC. This has translated into a decreased use of pre-treatment chemicals in drinking water treatment and cautious use of these chemicals in treatment ponds. For example, government agencies in Canada now recommend 50% lower additions of copper sulphate to drinking water reservoirs ; if such lower quantities are used the blue-green algae will still be controlled and some zooplankton will survive so that green algae blooms can be prevented.

Managing water treatment ponds will require the setting of clear objectives and the use of strategies that are most likely to achieve those objectives. The benefits of using Type 3 chemicals in any treatment pond management needs to be weighed against the potentially detrimental changes they may impart. A better knowledge-base need to be established for optimum management of man-made treatment ponds. This knowledge together with extensive small-scale interventions, such as those of introducing wetland species in artificial substrates will provide buffers against environmental pollution of lakes and rivers.

Acknowledgements

This study was carried out with assistance of the Green Municipal Enabling Fund which is financially supported by the Government of Canada and is administered by the Federation of Canadian Municipalities. Nothwithstanding such assistance, the views expressed are the personal views of the author, and the Federation of Canadian Municipalities and the Government of Canada accept no responsibility for them. In addition, the continued support from the City of Regina and its dedicated employees (Gary Nieminen, Ryan Johnston, Susan Olson, Wade Morrow and Taz Stuart) is gratefully acknowledged. Also, WateResearch Corp's Shannon Braithwaite and Norma Ruecker's assistance is much appreciated.

References

BASTIAN, R. K., P. E. SHANAGHAN, AND B. P. THOMPSON. 1989. Use of wetlands for municipal wastewater treatment and disposal-regulatory issues and EPA policies, p. 265-278. *In* D. A. Hammer [ed.], Constructed wetlands for wastewater treatment: municipal, industrial, and agricultural. Lewis Publishers, Chelsea, MI.

BRIX, H., AND H. H. SCHIERUP, 1989. The use of aquatic macrophytes in water pollution control. Ambio **18:** 100-107.

COOPER, P. F. 1993. The use of reed bed systems to treat domestic sewage: the European design and operations guidelines for reed bed treatment systems, p. 203-217. *In* G. A. Moshiri [ed.], Constructed wetlands for water quality improvement. Lewis Publishers, Boca Raton, FL.

KADLEC, R. H. 1995. Design models for nutrient removal in constructed wetlands, p. 173-184. *In* K. Steele [ed.], Animal waste and the land-water interface. Lewis Publishers, Boca Raton, FL.

_______, AND R. L. KNIGHT. 1996. Treatment wetlands. Lewis Publishers, Boca Raton, FL., 893p.

KNIGHT, R. L., R. W. RUBLE, R. H. KADLEC, AND S. REED. 1993. Wetlands for wastewater treatment: performance database, p. 35-38. *In* G. A. Moshiri [ed.], Constructed wetlands for water quality improvement. Lewis Publishers, Boca Raton, FL.

MITSCH, W. J., AND J. G. GOSSELINK. 1993. Wetlands. Academic Press, New York, NY. 722p.

PETERSON, H. G. 1998. Use of constructed wetlands to process agricultural wastewater. Can. J. Plant Sci. **78:** 199-210.

PETERSON, H. G., S. E. HRUDEY, I. A. CANTIN, T. R. PERLEY, AND S. L. KENEFICK. 1995. Physiological toxicity, cell membrane damage and the release of dissolved organic carbon and geosmin by *Aphanizomenon flos-aquae* after exposure to water treatment chemicals. Wat. Res. **29:** 1515-1523.

REED, S., R. CRITES, AND J. MIDDLEBROOKS. 1995. Natural systems for waste management and treatment, 2nd ed. McGraw-Hill, New York.

RYDING, S. O., AND W. RAST. 1989. The control of eutrophication of lakes and reservoirs. UNESCO, Paris, France. 314p.

STONE, J. A., AND D. E. LEGG, 1992. Agriculture and the Everglades. J. Soil Water Conservation **47:** 207-215.

TCHOBANOGLOUS, G. 1993. Constructed wetlands and aquatic plant systems: research, design, operational, and monitoring issues, p. 23-24. *In* G. A. Moshiri [ed.], Constructed wetlands for water quality improvement. Lewis Publishers, Boca Raton, FL.

WHIGHAM, D. F. 1995. The role of wetlands, ponds, and shallow lakes in improving water quality, p. 163-172. *In* K. Steele [ed.], Animal waste and the land-water interface. Lewis Publishers, Boca Raton, FL.

Chapter 4

Strategies for Ecological Modeling

4-1. Generic Numerical Models in Aquatic Ecology

Louis Legendre

Villefranche Oceanography Laboratory, BP 28, 06234 Villefranche-sur-Mer Cedex, France

Abstract

There are two broad approaches to numerical ecological models, i.e. specific and generic. In the specific approach, biological quantities in the model must be observed locally, which is often difficult or/and time-consuming. In contrast, the generic approach does not require local values of biological quantities in the model, these being computed with generic equations. In general, generic equations estimate the biological quantities in models from local values of easily observed environmental variables (e.g. temperature). Generic models have both advantages and drawbacks. On the one hand, there is no need to estimate locally biological quantities in the system, and the same model can be applied to homologous systems. On the other hand, generic models may fit less precisely local situations than specific models. In practice, the two approaches are often combined, e.g. in order to obtain rapidly at least preliminary answers to site-specific questions, generic models may be simplified to fit and explore the local situations. When detailed answers are required, specific models may be constructed using parts of existing generic models, which is an efficient use of knowledge developed by the international community, for resolving local problems.

Numerical Models in Ecology

Numerical models are used more and more in ecology. However, these models often appear quite mysterious to those who are not familiar with their development and use. Hence, our first question to is: What is a numerical ecological model?

Models are used in all scientific disciplines. They may be physical or numerical. In all cases, models are simplified representation of more complex, natural or artificial systems. In the past, most models were physical, e.g. small-scale models used to study the properties of planes or plane parts in wind tunnels. Biological models (e.g. white rats used to study the effects of drugs on people) are another type of physical model. Even if physical models are still very much in use, the rapidly increasing power of computers make numerical models more and more attractive.

Here is an example of the development in three steps of a very simple numerical ecological model. The three steps are: conceptual model, graphical box model and numerical model. Simple relationships among krill (small, shrimp-like crustaceans), whales and fishermen in the Southern Ocean can be described by the following sentences (conceptual model):

1. Whales eat krill
2. Fishermen catch both whales and krill

In other words, there is a natural food web that comprises krill and whales. Superimposed on this food web, fishermen are hunting whales, and also in some cases fishing krill.

These relationships can be represented in a box model (Fig. 1). In the Figure, boxes are ecological compartments, and arrows are relationships between two compartments. The boxes and arrows in Fig.1a can be replaced by equations with the following general forms (numerical model):

1. Krill biomass = *f* (Krill growth - Predation on krill by whales - Krill fishing mortality)
2. Whale biomass = *f* (Food assimilation by whales - Whale natural mortality - Whale fishing mortality)

In order to obtain the numerical model, these equations must be developed into expressions with variables and constants. This is beyond the purpose of the present paper.

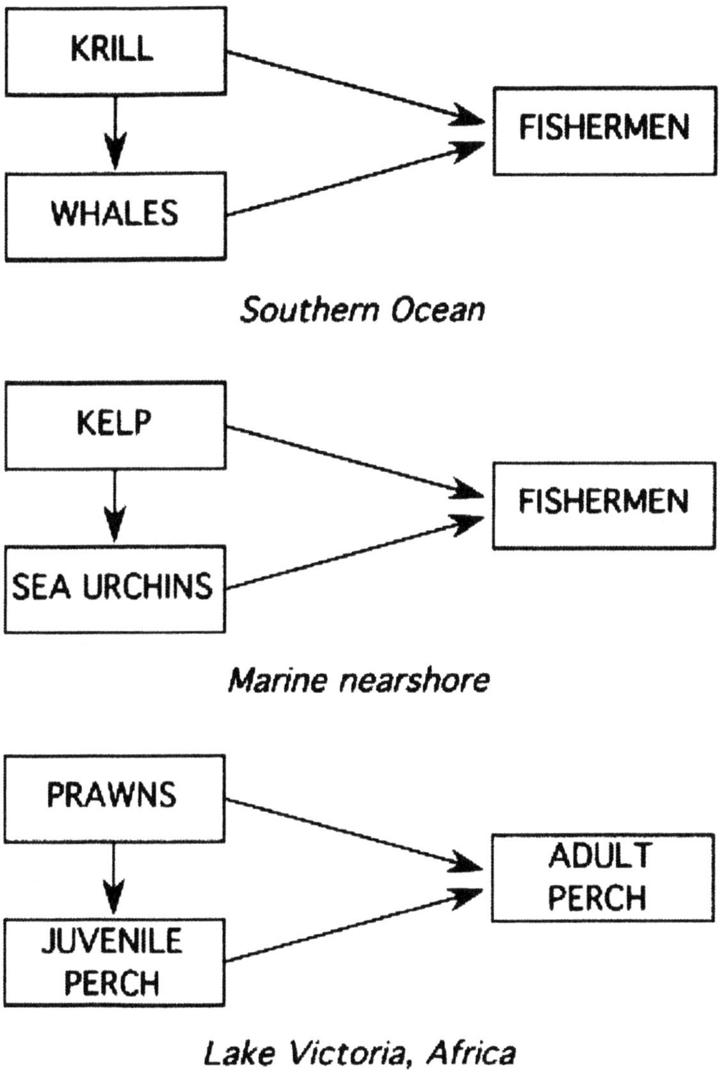

Fig. 1. Simple example of generic ecological model, applied to three homologous systems: (a) krill-whales-fishermen, in the Southern Ocean, (b) kelp-sea urchins-fishermen, in a marine nearshore environment and (c) prawns-juvenile perch-adult perch, in Lake Victoria, Africa (Witte et al. 1992).

Numerical ecological models can be used for various purposes. For example, the above equations could be used to calculate future abundances of whales, e.g. to estimate how many whales would be left in twenty years time if fishing on krill and whales continued as presently. Another use of such equations could be to study the roles of ecological variables in the krill-whales-fishermen system, e.g. what would happen to krill and whales if fishing on one, or the two groups increased, decreased, or stopped (whale hunting in the Southern Ocean is currently banned, but scientific harvesting continues, e.g. Hansom and Gordon 1998).

There are two broad approaches to ecological models, i.e. specific and generic. In the specific approach, the biological quantities used in the model must be observed locally. For example, in a specific version of the above simple model, the values for krill growth, predation on krill by whales, fishing mortality of krill and whales, etc. would need to be determined at sea. One serious problem is that obtaining such values is often difficult or/and time-consuming. In the generic approach, one does not need the local values of biological quantities used in the model, these being computed using generic equations that describe, for example, growth, natural mortality, the effects of fishing and predation, etc. In general, generic equations compute the biological quantities in models from local values of easily observed environmental variables. For example, bacterial respiration in the water column is rarely estimated directly, because of methodological difficulties. In order to nevertheless obtain estimates of bacterial respiration, Rivkin and Legendre (2001) proposed an equation that computes this quantity from temperature and bacterial production, these latter variables being much easier to obtain in the field than bacterial respiration.

Generic ecological models have advantages and drawbacks. Among the advantages is the fact that there is no need, in generic models, to estimate locally the biological quantities in the system. Another advantage is that the same generic model can be applied to different, homologous systems. For example, the equations developed to model the above simple krill-whales-fishermen system could also be used to model any system where (1) one predator exploits two resources, and (2) one of the resources exploits the other. Three such systems are illustrated in Fig. 1. In Figs. 1a and 1b, the predator is the fisherman, whereas in Fig. 1c, it is the adult perch; in Figs. 1a, 1b and 1c, krill are exploited by whales, kelp by sea urchins and prawns by juvenile perch, respectively. The same generic model (equations) could be used to study these widely different ecological situations. Among the drawbacks of generic models are the facts that they often fit less precisely local situations than specific models.

Examples of Generic Ecological Models

This section briefly discusses two examples of generic ecological models. The first illustrates the development of a model, from a general concept to its application, and the second shows the application to lakes of a generic model initially developed for the ocean.

Cousins (1980, 1985) proposed a new generic approach to ecosystems (valid for terrestrial and aquatic systems) that he called trophic continuum model. In that model, energy or matter flows among three functional trophic categories (autotrophs, heterotrophs and detritus) and between states within categories. In aquatic pelagic systems, the states within categories are size classes. Hence, this conceptual model provides an operational framework for the numerical modelling of flows in ecosystems, based on three functional trophic categories and size classes within categories.

Moloney and Field (1991) used the trophic continuum approach to build up a size-based model of the planktonic food web. The authors divided the autotrophs and heterotrophs into three and four size classes, respectively, and the detritus into dissolved (DOC) and particulate organic carbon. In their model, all transfer processes are size-dependent, and all parameters (e.g. production, respiration) are determined by body size, based on empirical relationships. The major carbon flows for phytoplankton include photosynthesis, exudation of DOC, respiration, senescence, and sinking; the flows for zooplankton include respiration, grazing, predation, egestion and sinking of faecal pellets. The model could easily accommodate different numbers of size classes, or different size ranges of organisms.

Moloney et al. (1991) used the model developed by Moloney and Field (1991) to simulate three contrasting planktonic food webs in the southern Benguela ecosystem (southeastern Atlantic Ocean): coastal stratified waters of the Agulhas Bank, a coastal upwelling area, and the oligotrophic oceanic waters of the southeastern Atlantic. In the three systems, a large proportion (70-90%) of the phytoplankton gross production is dissipated through autotrophic and heterotrophic respiration. Of the carbon reaching heterotrophs, the bulk (30-50%) occurs through grazing on phytoplankton, with only a small proportion (<10%) being taken up by bacteria and transferred to large heterotrophs through predation. The largest heterotrophs in the model (25-125 μm + 125-625 μm) graze only a very small proportion of phytoplankton production: 3.1% on the Agulhas Bank, 11.7% in the Benguela upwelling, and practically nothing in oceanic waters. The latter is because the bulk of oceanic phytoplankton production is <5 μm, so that it cannot be ingested by large heterotrophs. Even smaller proportions of phytoplankton production (<1%) reach the large heterotrophs through predation. The

Table 1. Fate of phytoplankton production (P) in temperate lakes during the spring and autumn blooms, and summer: fractions of P respired (R) and consumed (C) in the upper water column, and exported to depth (D), computed using the equations and empirical constants of Legendre and Rassoulzadegan (1996). The input variables are the proportion of large-sized (L) in total P (L/P), and the matching (M) between P and its grazing by zooplankton (values of L/P and M range between 0 and 1, and are identical in the upper and lower Tables). Observed values (upper Table) and model results (lower Table), from Legendre (1999).

Lake condition	L/P	M	R/P	C/P	D/P
Spring bloom	*0.85-0.95*	*0.1*	0.1	0.0-0.2	0.8-0.9
Autumn bloom	*0.85*	*0.5*	0.3	0.2	0.5
Summer	*0.50-0.85*	*0.9-1.0*	0.2-0.4	0.3-0.7	0.1-0.5

Lake condition	L/P	M	R/P	C/P	D/P
Spring bloom	*0.85-0.95*	*0.1*	0.1-0.2	0.1	0.8-0.9
Autumn bloom	*0.85*	*0.5*	0.1	0.3	0.6
Summer	*0.50-0.85*	*0.9-1.0*	0.1-0.4	0.4-0.5	0.2-0.4

microbial food web is generally inefficient as a carbon pathway to large heterotrophs. In oceanic waters, however, the microbial food web is the only available carbon pathway toward large heterotrophs, so that it plays a significant role there. In coastal waters, the microbial food web is a small source of carbon for the large heterotrophs.

The second example discusses the application to lakes of a generic model initially developed for the ocean. Legendre and Rassoulzadegan (1996) proposed a numerical model to estimate the fate of marine phytoplankton production (P). Their equations can be used to compute the fractions of P that are respired (R) and consumed (C) in the upper water column, and exported to depth (D). In order to compute these quantities with the model, one needs two empirical constants, and two input variables. The latter are the proportion of large-sized (L) in total P (L/P), and the matching (M) between P and its grazing by zooplankton (values of the two input variables range between 0 and 1). M = 1 corresponds to situation where there is tight spatio-temporal coupling between P and grazing, and M = 0 to situations where the two processes are decoupled. Legendre (1999) applied the model to three temperate lakes (Lake Biwa, Japan, Lake Kinneret, Israel, and Lake Constance, Germany). The observed and computed values of R/P, C/P and D/P are summarized in Table 1 for three conditions in the lakes studied, i.e. spring and autumn blooms, and summer. Comparing the observed and computed values shows good agreement between the observations and model results.

Conclusions

Numerical models are essential tools in ecological research. They can be used to study the roles of variables in ecosystems, and explore scenarios of future conditions. There are two broad approaches to ecological models, i.e. specific and generic. Among the advantages of generic models are the facts that one does not need to estimate locally the biological quantities in the system, and the models can be applied to homologous systems. Among the drawbacks are that generic models may fit less precisely local situations than specific models. In practice, the two approaches are often combined. For example, in order to obtain rapidly at least preliminary answers to site-specific questions, generic models may be simplified to fit and explore the local situations. Alternatively, when detailed answers are required, specific models may be constructed using parts of existing generic models, which is an efficient use of knowledge developed by the international community, for resolving local problems.

References

COUSINS, S. H. 1980. A trophic continuum derived from plant structure, animal size and a detritus cascade. J. Theor. Biol. **82:** 607-618.

________. 1985. The trophic continuum in marine ecosystems: structure and equations for a predictive model, *In* R. E. Ulanowicz and T. Platt [eds.], Ecosystem theory for biological oceanography. Can. Bull. Fish. Aquat. Sci. **213:** 76-93.

HANSOM, J. D., AND J. E. GORDON. 1998. Antarctic environments and resources: A geographical perspective. Longham.

LEGENDRE, L. 1999. Environmental fate of biogenic carbon in lakes. Jpn. J. Limnol. **60:** 1-10.

________, AND F. RASSOULZADEGAN. 1996. Food-web mediated export of biogenic carbon in oceans: environmental control. Mar. Ecol. Prog. Ser. **145:** 179-193.

MOLONEY, C. L., AND J. G. FIELD. 1991. The size-based dynamics of plankton food webd. I. A simulation model of carbon and nitrogen flows. J. Plankton Res. **13:** 1003-1038.

________, ________, AND M. I. LUCAS. 1991. The size-based dynamics of plankton food webd. II. Simulations of three contrasting southern Benguela food webs. J. Plankton Res. **13:** 1039-1092.

RIVKIN, R., AND L. LEGENDRE. 2001. Biogenic carbon cycling in the upper ocean: effects of microbial respiration. Science **291:** 2398-2400.

WITTE, F., T. GOLDSCHMIDT, P. C. GOUDSWAARD, W. LIGTVOET, M. J. P. VAN OIJEN, AND J. H. WANINK. 1992. Species extinction and concomitant ecological changes in Lake Victoria. Neth. J. Zool. **42:** 214-232.

4-2. Site-Specific Models and the Importance of Benthic-Pelagic Coupling

Masumi Yamamuro

Institute for Marine Resources and Environment, Geological Survey of Japan, AIST Tsukuba Central 7, Tsukuba 305-8567 Japan

Abstract

The purpose of applying ecological models to lakes is to improve their management and protection, and to predict future changes in the lake water environment. Although ecological models cannot necessarily incorporate all significant processes, they should integrate the biomass and metabolic characteristics of the dominant primary producers, as well as the key secondary producers which directly graze upon them. In deep water systems, phytoplankton dominates the primary production, and zooplankton represents the secondary producer level. The biomass and metabolism of these two components are often described as a function of environmental variables based on the previous studies elsewhere. However, in shallow water lakes and even the inshore areas of deeper lakes, benthic organisms such as filter-feeding bivalves can play an important role as grazers on the phytoplankton. This type of local, site-specific effect is essential to incorporate within whole-lake models. This additional component may complicate the model, although can sometimes lead to useful simplifications. The development of models for specific sites requires a clear formulation of the objectives, and the acquisition of appropriate, site-specific data. Field observations are therefore essential to guide the modeling effort, to identify critical, site-specific mechanisms and to obtain realistic estimates of the input parameters for the model.

Introduction

There are generally two main objectives when ecological models are applied to lakes. One objective is to improve the ways and directions for management and protection of lakes. We can change the parameters in the models under different conditions, and run several cases to find a suitable method for improved lake water management and protection. Another objective is to predict future changes in lakes. As it is not easy to predict the future and the result may sometimes bring serious concerns to the public, this objective requires models based on a reliable scientific understanding of how the ecosystem operates. If the model is too general and deviates far from local conditions, it loses all value as a tool for guiding and implementing future management strategies. With regard to this, site-specific models have advantages over a more generic approach. The present chapter illustrates the importance of this local perspective by way of examples from Lakes Nakaumi and Shinji in the southwestern part of Honshu Island, Japan.

Effects of Benthic Macro-Invertebrates on Water Quality

Figure 1 shows a biochemical model used for computing nutrient flow in Lake Nakaumi. This may look very complicated, but each component is formulated in generic terms and each state variable and flux is calculated not with observed values, but with generic equations. For example, the factors concerning zooplankton (B6 – B10) can be calculated with the general equations as shown in Table 1.

L. Nakaumi is a polyhaline coastal lagoon with 5.4 m mean depth and 17.1 m maximum depth. In 1963, a reclamation project to obtain a new rice paddy in this lake started, and 23.4% (=652 ha) of the lake became land by 1981. Another 1417 ha at north west of the lake was planned to be drained, but this project was cancelled in 2000. One reason for the cancellation was the decreased demand for new land for rice paddies. Another was the possibility that the reclamation could accelerate the deterioration of water quality in the lake. The lake is already eutrophic with surface water total-N values around 600 μ g/l and total-P values around 60 μ g/l, and there was concern about further degradation.

In Lake Nakaumi, *Musculus senhausia,* a filter-feeding bivalve, is very abundant

Biochemical processes concerned in the coastal ecological model

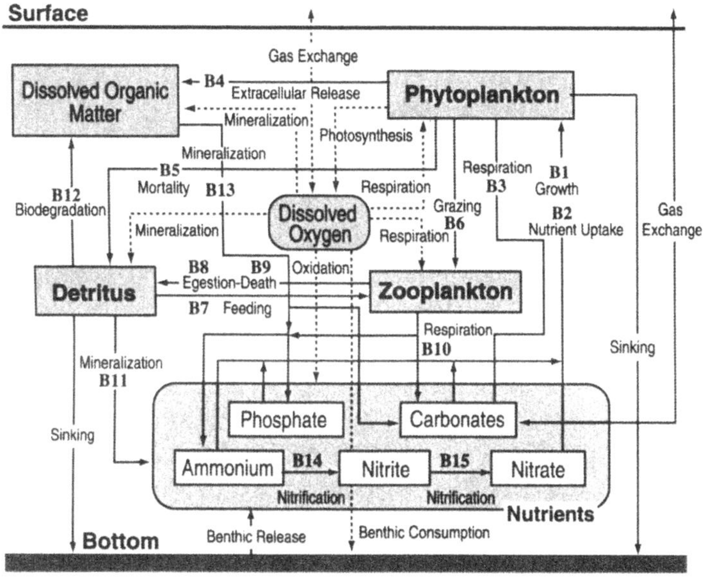

Fig. 1. Biochemical processes incorporated in the coastal ecological model (From Nakata et al. 2000).

Table 1 Formulation of biochemical processes as influenced by zooplankton grazing. B-number shows the position of each process in Fig.1 (from Nakata et al. 2000).

Formulation of biochemical processes

	Biochemical process	Formulation	Parameters
Zooplankton			
B6	Grazing	$B_6 = \dfrac{P}{P+\mathrm{POC}}\nu_5(P, \mathrm{POC})Z$	
B7	Detrital feeding	$B_7 = \dfrac{\mathrm{POC}}{P+\mathrm{POC}}\nu_5(P, \mathrm{POC})Z$	
	Total ration	$\nu_5 = R - \exp(\beta_{R_{max}} T)\,\mu_4(P, \mathrm{POC})$	R_{max}, $\beta_{R_{max}}$
	Food limitation	$\mu_4 = 1 - \exp\{\lambda[\Pi^* - (P+\mathrm{POC})]\}$	λ, Π^*
B8	Egestion	$B_8 = (1-e)(B_6 + B_7)$	e
B9	Natural mortality	$B_9 = \nu_6(T, Z)Z$	
	Mortality rate	$\nu_6 = Z_{mot}\exp(\beta_{Z_{mot}} T)Z$	Z_{mot}, $\beta_{Z_{mot}}$
B10	Respiration	$B_{10} = \nu_7(T, P, \mathrm{POC})Z$	
	Respiration rate	$\nu_7 = Z_{resp}\exp(\beta_{Z_{resp}} T) + \eta\nu_5(T, P, \mathrm{POC})$	Z_{resp}, $\beta_{Z_{resp}}$, η

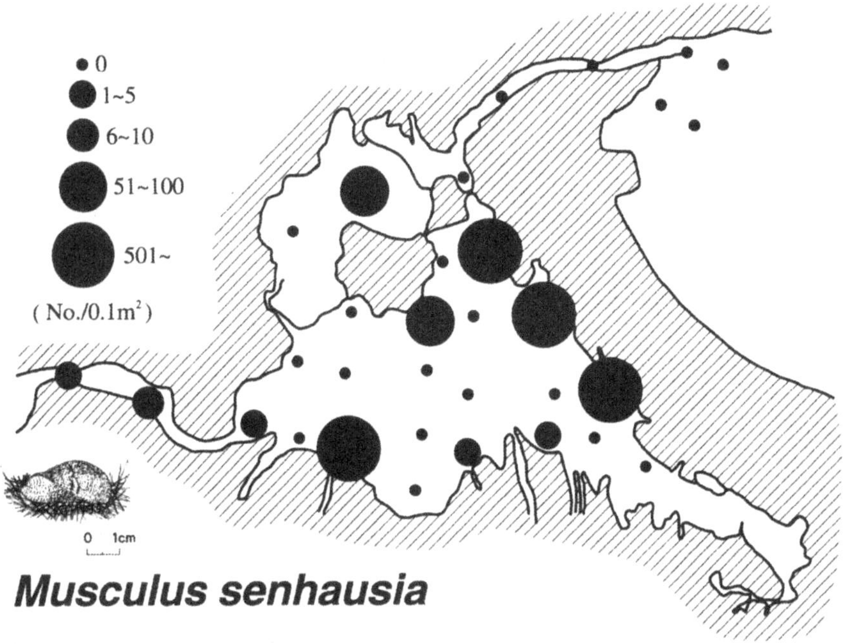

Fig. 2. Distribution of *Musculus senhausia* in Lake Nakaumi.

(Fig. 2). They cluster together by secreted byssal threads and cover wide expanses of sediment. Since the secondary producers, which directly graze the phytoplankton, control the amount of phytoplankton through supplying the nutrient at the same time they graze them (Fig. 3), it is likely that the model without mussels created results that deviated far from the reality of actual nutrient fluxes. None of the previous ecological models used to examine the effects of reclamation on water quality included a macrobenthos component in the ecosystem.

To obtain the appropriate input data for a coupled pelagic-benthos model we made observations on *M. senhausia* biomass and its seasonal change in the lake (Yamamuro et al. 2000), and also determined the metabolism of *M. senhausia* in laboratory experiments (Inoue and Yamamuro 2000). These results were subsequently incorporated into the ecological model, as shown in Table 2.

As shown in Fig. 4, the results of numerical simulation with the coupled ecological model agreed well with the observed values. We then compared the results for phytoplankton stocks, especially critical to lake management, between the simulations with and without the forcing through *M. senhausia*. In August, the simulation with the

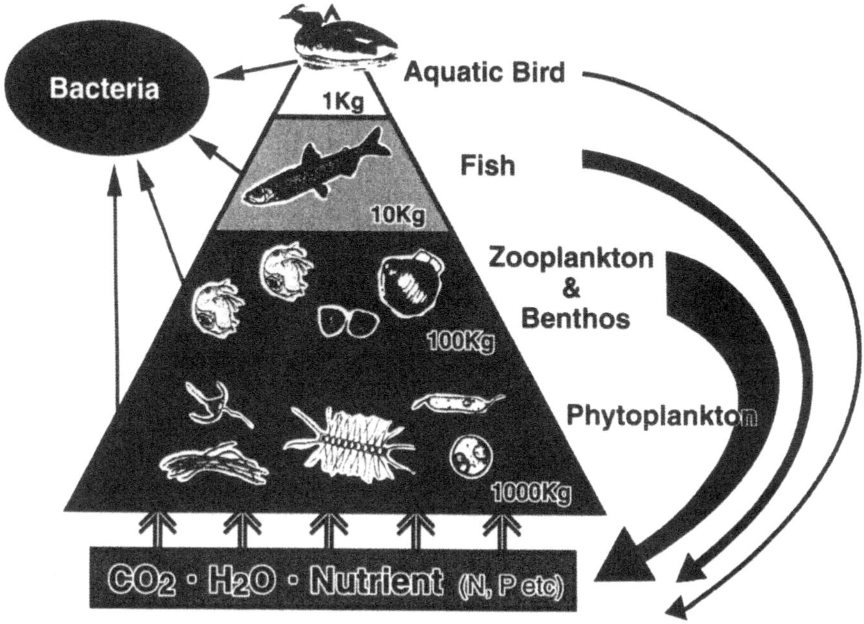

Fig. 3. Schematic picture of nutrient cycle in lakes highlighting the food chain through macrosized organisms.

Table 2 Parameter values related to *Musculus senhausia* metabolism (From Nakata et al. 2000).

Parameter	Unit	*M. senhausia*
Filtration rate at 0℃	(g dry weight h)$^{-1}$	0.495
Temperature coefficient for filtration	1/℃	0.1082
Maximum filtration rate	(g dry weight h)$^{-1}$	8.73
Respiration rate at 0℃	ml(kg dry weight h)$^{-1}$	193.4
Temperature coefficient for respiration	1/℃	0.1174
O/P ratio	by weight	205.8
O/N ratio	by weight	20.6

forcing through *M. senhausia* showed the stock of phytoplankton as 39 tons (Fig. 5a). Without it the stock became 67 tons (Fig. 5b), a 186% overestimate. Therefore, the effect of *M. senhausia* cannot be neglected in the ecological model of Lake Nakaumi, and *M. senhausia* clearly exerts a strong pressure resulting in much lower

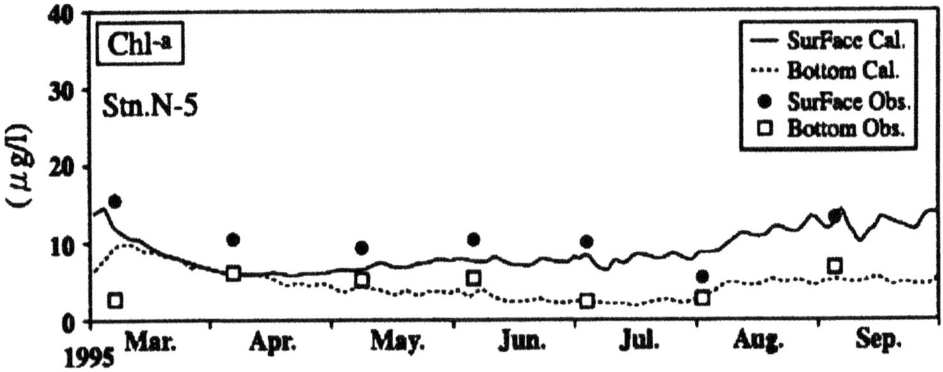

Fig. 4. The simulated chlorophyll-a concentrations in comparison with data from monthly measurements in Lake Nakaumi (From Nakata et al. 2000).

phytoplankton biomass concentrations and higher lakewater transparencies than would otherwise be the case.

The effects of filter-feeding bivalves on water quality have been incorporated into ecological models for estuaries (e.g., Gerritsen et al. 1994) and a coastal lagoon (Klepper et al. 1994). This increased awareness of the importance of benthic grazers in brackish water habitats may be because the number of grazing species in these environment is far lower than fresh and marine water (Remane 1971), and filter-feeding bivalves often predominate in terms of both production and biomass (Wolff 1983).

In freshwaters, a single species of filter-feeding bivalves may not be significant in term of biomass, but the overall benthic community of all filter-feeding species may be critically important. Some freshwater bivalves such as the Asiatic clam (*Corbicula*) and zebra mussel, are being widely dispersed by human activities (e.g. ship ballast waters) and these exotic species often show spectacular population increases after invasion, for example in North American lakes (Ricciardi et al. 1995). Information about the filtration characteristics of these animals' populations (e.g., Holland 1993) is essential for applying numerical simulation to determine water quality variables such as algal biomass and transparency in such lakes.

The Importance of Field Observations

Obtaining the biomass and the metabolic factors for each bivalve is often difficult and time consuming. Therefore, the acquisition of metabolic data in local studies is of great help for the simulation in other areas where such observations have not been performed.

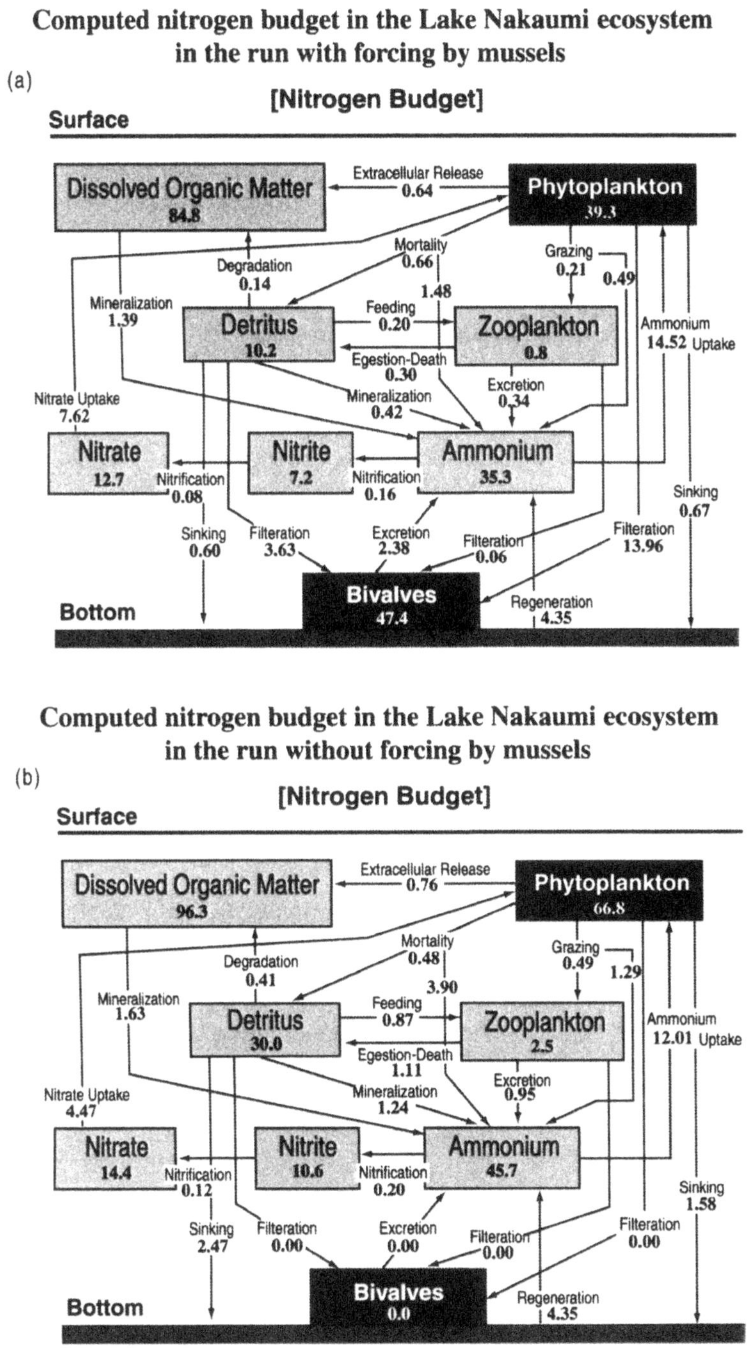

Fig. 5. Integrated nitrogen cycle in Lake Nakaumi ecosystem (a) in the standard run and (b) run without forcing by bivalves. Units: flow (ton/day) and stock (ton) (From Nakata et al. 2000).

However, biomass data in each location cannot be estimated from the data of other locations. Field surveys of biomass data are essential to tune the ecological model to the local setting. For example, we observed the carnivorous mollusk species eating *M. senhausia* in shallow aerobic bottom waters of L. Nakaumi during the field survey. Some crustaceans, such as crabs and shrimps, also ate *M. senhausia* in an aquarium (Yamamuro unpubl.). This means that *M. senhausia* can expand their distribution when bottom becomes more aerobic, but such expansions may be limited by subsequent predation. Thus, when changes occur in the flora and fauna as a result of human actions or natural forcing, existing models may no longer yield adequate prediction of the present or future states of the lake environment. In case of Lake Nakaumi, some citizens wish to use the existing ecological model (Fig. 1) to predict the impact on water quality of removing dikes from the lake, allowing the bottom waters to become more aerobic. Without field observations, the model results imply a wider distribution and biomass of *M. senhausia* under more aerobic conditions. Based on the field observation, we know that the present model is inadequate to address this question and must be modified to incorporate higher level trophic interactions.

Specific models therefore requires careful scrutiny, relatively new information, and understanding derived from the field. This does not necessarily mean the model must become more complex. Rather it is important to identify and extract the key processes that control the modeled water quality variables, and to suppress the less important subcomponents.

Lake Shinji is an oligohaline coastal lagoon that connects to L. Nakaumi through the Ohashi River. The Asiatic clam, *Corbicula japonica*, is the dominant filter-feeding bivalve in this lake. Before the reclamation of northwest Lake Nakumi was completed, the brackish waters of this lake were planned to be replaced with freshwater for rice paddy agriculture (Swinbanks 1988). It was argued that the water quality of Lake Shinji would greatly improve after freshening because there would no longer be salinity stratification and wind-induced mixing would prevent or lessen the frequency of anaerobic conditions. Yamamuro and Koike (1993) proposed that the uptake of particulate of organic matter (POM) by *C. japonica* accounts for half of the primary production in this lake during summer, which suggests that freshening would in fact lead to increased algal biomass and reduced water quality given that *C. japonica* cannot reproduce in freshwater and would become extinct in the modified lake.

The distribution of *C. japonica* is restricted to depths shallower than 3m, which account for only one third of the total lake area. Because the amount of POM entering the lake from the freshwater Hii River is negligible, some unknown factor was

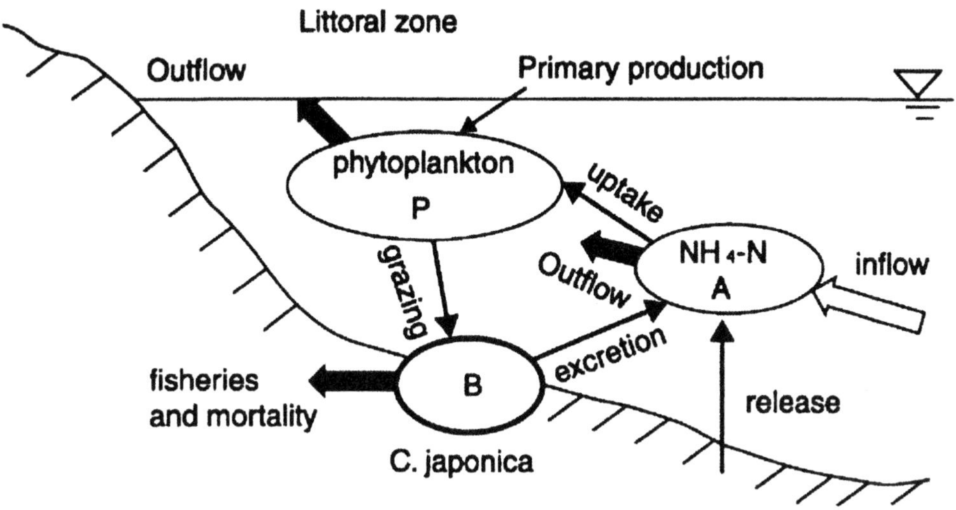

Fig. 6. Model variables and processes taken into account in the mathematical model. Nakamura and Kerciku (2000).

suspected which enabled the benthic-pelagic coupling as a means of supplying POM from the whole pelagic zone to shallow bottom where *C. japonica* inhabits.

Nakamura and Kerciku (2000) conducted the field observations in Lake Shinji in summer, and found that the water column was thermally homogeneous at night. Thus overnight heat loss from the water surface appears to be sufficient to cause mixing by penetrative convection throughout the water column, thereby enabling *C. japonica* to capture phytoplankton that has grown in the surface pelagic zone during diurnal stratification.

With this information, Nakamura and Kerciku (2000) made a simple mathematical model to predict the chlorophyll-a concentration (Fig. 6). The parameters used are listed in Table 3.

As shown in Fig. 7, simple mathematical model with small number of parameters gave a fairly close fit to the observed data. These results underscore the critical importance of *C. japonica* in controlling algal biomass, and also suggests that other, less important components need not be incorporated in the model, or need to be less precisely defined. The key to establishing successful ecological models for addressing specific water quality problems is to define clearly the objectives of the simulation and to modify generic models to take into account new insights that are derived from detailed observations in the field. Recent advances in technologies for field observation

Table 3 List of model parameters.

P_{max}	primary production rate constant	0.2 day^{-1} at 20°C
K_m	half saturation constant for NH_4-N	0.014 g m^{-3}
k_d	mortality of phytoplankton	0.08 day^{-1}
γ	conversion factor for weight	0.0223 g dry weight g wwt^{-1}
F	filtration rate	0.14 m g dry weight $^{-1}$ day^{-1} at 20°C
E	excretion rate of NH_4-N	0.003 g N g dry weight $^{-1}$ day^{-1} at 20°C
k_b	loss rate by fisheries and mortality	0.003 day^{-1}
α	conversion constant	6.3 g N g $chl.a^{-1}$
β_1	assimilation efficiency for N	0.45
β_2	conversion constant	4.6×10^2 g wwt g N^{-1}
R	NH_4-N release rate from sediment	0.04 g m^{-2} day^{-1}
A_{in}	influent NH_4-N concentration	0.0035 g m^{-3}
D_x	horizontal diffusion coefficient	8.6×10^4 m^2 day^{-1}
τ	hydraulic retention time	103 days

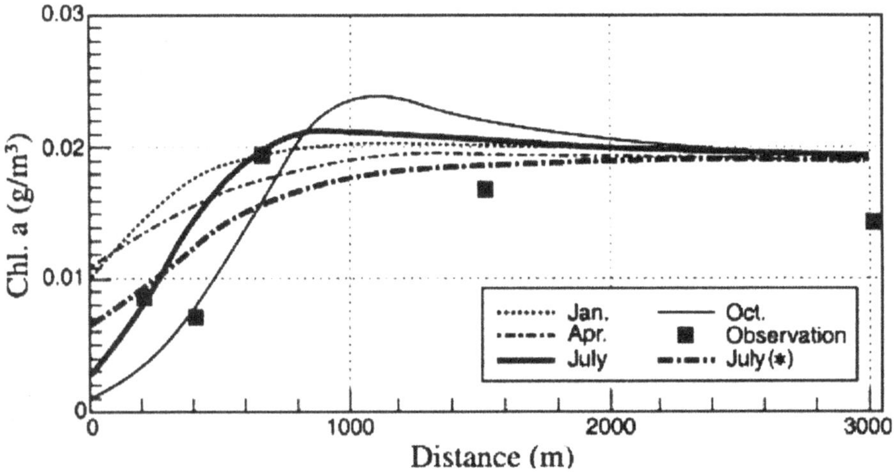

Fig. 7. Calculated results of chlorophyll-a concentration. Symbols denote observations averaged over for all data at each station.

such as ROVs, AUVs and remote sensing, provide considerable promise for an improved level of obtaining such data. In the lakes described here, as in other shallow water ecosystems (Nixon 1988), field observations highlighted the need to consider benthic-pelagic coupling as a key ecological process, and the importance of zoobenthos data for adequate simulations. Our results show that water quality models cannot be formulated in the absence of a field program to obtain the necessary input data and to identify those processes which exert a dominant influence in the local ecosystem.

References

GERRITSEN, J., A. F. HOLLAND, AND D. E. IRVINE. 1994. Suspension-feeding bivalves and the fate of primary production: An estuarine model applied to Chesapeake Bay. Estuaries **17:** 403-416.

HOLLAND, R. E. 1993. Changes in planktonic diatoms and water transparency in Hatchery Bay, Bass Island Area, Western Lake Erie since the establishment of the zebra mussel. J. Great Lakes Res. **19:** 617-624.

INOUE, T., AND M. YAMAMURO. 2000. Respiration and ingestion rates of the filter-feeding bivalve *Musculista senhousia*: implications for water-quality control. J. Marine Syst. **26:** 183-192.

KLEPPER, O., M. W. M. VAN DER TOL, H. SCHOLTEN, AND P. M. J. HERMAN. 1994. SMOES: a simulation model for the Oosterschelde ecosystem. Hydrobiologia **282/283:** 437-451.

NAKAMURA, Y., AND F. KERCIKU. 2000. Effects of filter-feeding bivalves on the distribution of water quality and nutrient cycling in a eutrhophic coastal lagoon. J. Marine Syst. **26:** 209-221.

NAKATA, K., F. HORIGUCHI, AND M. YAMAMURO. 2000. Model study of Lakes Shinji and Nakaumi --- a coupled coastal lagoon system. J. Marine Syst. **26:** 145-169.

NIXON, S. W. 1988. Physical energy inputs and the comparative ecology of lake and marine ecosystems. Limnol. Oceanogr. **33:** 1005-1025.

REMANE, A. 1971. Ecology of brackish water. p. 2-210. *In* A. Remane and C. Schlieper [eds.], Biology of brackish water. John Wiley & Sons.

RICCIARDI, A., F. G. WHORISKEY, AND J. B. RASMUSSEN. 1995. Predicting the intensity and impact of *Dreissena* infestation on native unionid bivalves from *Dreissena* field density. Can. J. Fish. Aquat. Sci. **52**: 1449-1461.

SWINBANKS, D. 1988. Plans for giant freshwaterpond opposed by Japansese. Nature **332:** 480.

WOLFF, W. J. 1983. Estuarine benthos. p. 151-182. *In* B. H. Ketchum [ed.], Estuaries and Enclosed Seas. Elsevier.

YAMAMURO, M., AND I. KOIKE. 1993. Nitrogen metabolism of the filter-feeding bivalve

Corbicula japonica and its significance in primary production at a brackish lake in Japan. Limnol. Oceanogr. **38:** 997-1007.

________, J. HIRATSUKA, AND Y. ISHITOBI. 2000. Seasonal change in a filter-feeding bivalve *Musculista senhousia* population of a eutrophic estuarine lagoon. J. Marine Syst. **26:** 117-126.

Chapter 5

Recovery from Eutrophication

5-1. Restoration of Eutrophic Lakes: a Global Perspective

Erik Jeppesen[1,2], Martin Søndergaard[1], Jens Peder Jensen[1] and Torben L. Lauridsen[1]

[1]*National Environmental Research Institute, Dept. of Freshwater Ecology, P.O. Box 314, Vejlsøvej 25, DK-8600, Silkeborg, Denmark (ej@dmu.dk),*

[2]*Dept. of Botanical Ecology, University of Aarhus, Nordlandsvej 63, DK-8230, Risskov, Denmark*

Abstract

Many lakes world-wide suffer from eutrophication as a result of high external nutrient inputs from domestic sewage, industry and agricultural activities. Increased demands for water by a growing and developing human population as well as enhanced global warming may escalate the eutrophication on a global scale in the next century. Yet, in some countries large efforts are now being made to combat eutrophication by reducing the phosphorus input. Some lakes respond rapidly to such loading reductions, while others are highly resistant due to high internal phosphorus loading (chemical resistance) or homeostatic effects of the food web altered by eutrophication (biological resistance). Some general models have been developed for the response of lakes to reduced loading, but major advances within this field can be expected in the future when more case studies appear. While these models may be used as a core for evaluating response patterns, local factors should always be considered to avoid wrong and often expensive decisions. To precipitate recovery from chemical and biological resistance, several physico-chemical and biological restoration methods have been developed. The biological methods include removal of planktivorous and benthivorous fish, stocking of piscivorous fish, protection or planting of submerged macrophytes, introduction of artificial structures, or addition of mussels. A widely applied method is removal of planktivorous and benthivorous fish. In many cases such efforts have yielded major improvements in water quality and the ecological state of the lakes. Yet, the listed restoration methods have mainly been applied to northern temperate lakes and cannot readily be transferred to subtropical and tropical lakes where the eutrophication-related problems are going to be greatest in the future. There is thus a major need for development and adaptation of methods focusing on south temperate, subtropical and tropical lakes.

Introduction

Many lakes suffer from increasing eutrophication due to high external loading of nutrients (Wetzel 1990), resulting in high algal biomass often accompanied by massive summer blooming of cyanobacteria or green algae, few submerged macrophytes, dominance of plankti-benthivorous fish and low water clarity (Moss 1998). Palaeoecological studies of lake sediments have shown that anthropogenic activities such as afforestation and early agriculture have long affected the nutrient state of lakes and their ecology. However, during the past century increased urbanization and sewage disposal, regulation of wetlands and streams and more intensive farming practices have greatly increased the nutrient loading to many shallow lakes world-wide. With an increasing human population (at present of 1.3% per year, UN 1998), a rising demand for water for agriculture, industry and humans (6-fold increase in consumption from 1990-1995, UN 1998) along with enhanced global warming, it is expected that eutrophication of lakes will increase substantially in the next century. Yet, in Europe, North America and other industrial countries, the nutrient loading from sewage and industry sources has been reduced significantly since the 1970s. However, considering the continuing significant loading from agriculture and scattered settlements, eutrophication is likely to remain a major problem even in these geographical areas.

While the eutrophication process has been studied extensively, comparatively few studies exist on the response of lakes to reduced loading, and the majority of these have focused on changes in chlorophyll *a* or the abundance and community structure of phytoplankton (Olsén and Willén 1980; Edmondson and Lehman 1981; Bernhardt et al. 1985; Marsden 1989; Sas 1989). Only few studies include the higher trophic levels in the pelagial (zooplankton and fish) or the microbial loop (Gaedke 1998). This chapter will provide a brief survey of the response of lakes to reduced loading and present a few case studies. We will also discuss the possibilities for encouraging improvement of the environmental state by interventions in the biological system (termed biomanipulation).

Response to Reduced Loading in Northern Temperate Lakes

While a number of general models have been developed for the response of lakes to increased loading, both as to changes in lake water nutrient levels, abundance of algae as chlorophyll *a* or biomass, and for changes in the composition of bottom invertebrates and fish (Vollenweider 1976; OECD 1982; Hanson and Leggett 1982; Hanson and Peters 1984; Reynolds 1984; Jeppesen et al. 1997, 2000), only few studies have endeavoured to establish general conceptual or empirical models for the recovery phase after nutrient loading reductions.

A notable exception is the study by Sas (1989) who compiled data from 18 well-studied lakes in Europe. He argued for a four-stage response of phytoplankton to nutrient loading reduction. At the first stage, no response is found, as the phosphate concentration is too high throughout the growing season to limit phytoplankton growth. At the second stage, P limitation occurs during part of the summer, leading to lower phytoplankton biomass per unit of volume but only small or no changes in biomass per unit area, due to a deepening of the photic zone because of the improved water clarity. At the third stage, biomass per unit of area is affected and, at the fourth stage, also changes in phytoplankton community composition occur. Most recent studies do not fully support this conceptual model. Results from Danish lakes suggest that the biomass and structure of phytoplankton and fish responded to an in-lake TP reduction even at high phosphate concentrations (Box 1).

Box 1. Example of response of 18 Danish lakes following a reduction in external phosphorus loading

The response of phytoplankton, zooplankton and fish to reduced nutrient concentrations following a TP loading reduction was studied in 18 Danish lakes. For the majority of lakes, a fast response to the TP reduction (but not necessarily to TP loading reduction due to internal loading) was found for phytoplankton (chlorophyll *a*) and fish, while less pronounced changes were observed for the zooplankton. In contrast to zooplankton biomass, phytoplankton and fish biomass generally declined, leading to an overall higher ratio of zooplankton to phytoplankton, which suggests enhanced grazing pressure on algae. Also a response at community level was found. Generally, the share of non-heterocystous cyanobacteria declined substantially, while the contribution of particularly heterocystous cyanobacteria, and in some lakes also dinophytes, cryptophytes and chlorophytes, increased. The biomass of planktivorous fish declined and the share of potential piscivores increased. Accordingly, the share of *Daphnia* spp. among cladocerans and the body weight of *Daphnia* spp. and cladocerans often increased, suggesting a reduced predation pressure on zooplankton. The response was not restricted to low-TP lakes, but was also found in lakes in the TP range 0.2-0.4 mg P l^{-1}. However, only minor changes to a TP reduction from 1.2 to 0.4 mg P l^{-1} were recorded in the most hypertrophic lake.

(Sources: Jeppesen et al. 2002, Søndergaard et al. 2002)

Several studies have shown a rapid response of phytoplankton; a classical one being the recovery study of Lake Washington (Edmondson and Lehman 1981) in which chlorophyll *a* declined concurrently with a reduction in TP following sewage abatement. Other examples are the Wahnbachtalsperre Reservoir, Germany, where only a one-year delay occurred in the response of chlorophyll *a* to a major decline in TP (Bernhardt et al. 1985) and studies in a number of temperate Wisconsin lakes, USA (Lathrop 1990), Lake Constance, Germany (Box 2), and sub-arctic lakes (Choulik and Moore 1992). A similar observation was recently made by Köhler et al. (2000) in Lake Müggelsee, Germany. Pronounced response to lake TP reductions in direction of more oligotrophic taxa has been found in many published re-oligotrophication studies from both deep and shallow lakes (e.g. Olsén and Willén 1980; Willén 1984, 2001; Wojciechowski et al. 1988; Edmondson and Lehman 1981; Gaedke 1998; Cronberg 1999; Beklioglu et al. 1999; Köhler et al. 2000; Box 1), though a delay in the response at community level, as suggested by Sas (1989), has been observed in some case studies (Polli and Simona 1992; Ruggio et al. 1998; Anneville and Pelletier 2000).

Box 2. Example of fast response to external phosphorus loading reduction – Lake Constance, Germany

Upper Lake Constance is large (surface area 472 km^2), deep (max depth 253 m, mean depth 101 m) and has a large catchment area (10,500 km^2) of which more than 70% are alpine mountains. Compared to its size the retention time is relatively short (4.2 y). During the 20th century the number of inhabitants in the catchment more than doubled (from 0.5 to 1.2 mill.). From 1955 to 1977 the TP loading increased substantially due to enhanced sewage release from cities and intensified agricultural practices, reaching a value 15-20-fold higher than that estimated for the turn of the century (Fig. 1). However, owing mainly to improved treatment of sewage water and substitution of P in detergents the loading declined substantially to the 1955 level during the next two decades. During the eutrophication phase phytoplankton biomass increased markedly (Fig. 1). Diatoms became dominant and within this group major changes occurred towards dominance of eutrophication-indicating species such as *Stephanodiscus* spp. As judged from the commercial catches of fish, fish biomass increased as well, but changed from dominance of whitefish (*Coregonus laveretus*) to dominance of species characteristic of more mesotrophic-eutrophic waters such as perch (*Perca fluviatilis*). A fast response to loading reduction occurred. The TP concentration started to decline less than five years after the loading reduction, from approx. 80 µg l^{-1} in the late 1970s to 20 µg l^{-1} in 1996. Additionally, the abundance of phytoplankton declined, diatoms lost their prevalence during summer, while the proportion of cryptophytes and chrysophytes increased. Also the fish stock responded, the biomass of catches decreased and whitefish again became dominant. Despite its large size and depth, Lake Constance is thus an illustrative example of a fast response to measures taken to reduce nutrient loading.

(Sources: Bäuerle and Gaedke 1998, Güde et al. 1998, Kümmerlin 1998, Eckmann and Rösch 1998)

Box 2. Continued

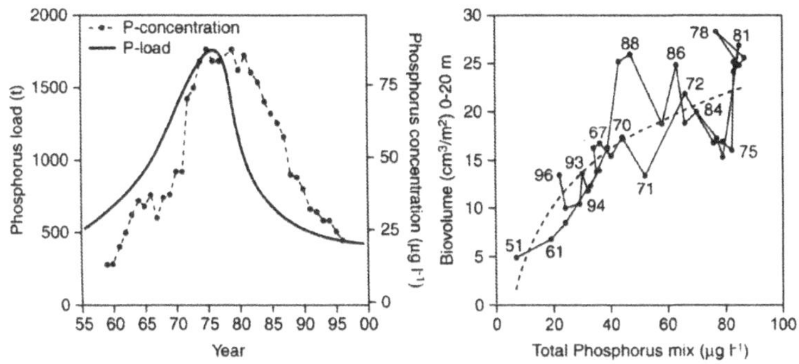

Fig. 1 a

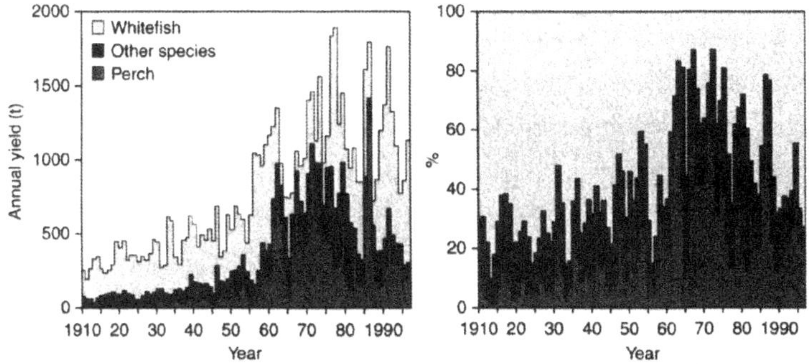

Fig. 1 b

Fig. 1. Top: Calculated external loading and total phosphorus concentration (left) and the relationship between algal biovolume and concentration of total phosphorus (right) in the epilimnion of Lake Constance during 1955-1979. Below: Annual catches (left) and relative composition of fish catches by commercial fishermen in Lake Constance during 1910-1997.

Relatively few studies exist on the effects of re-oligotrophication on fish, and they, unfortunately, mainly comprise analyses of changes in fish catches by anglers and commercial fishermen. Yet the studies published so far indicate that when the TP concentration in the lakes declines (not necessarily simultaneously with the loading reduction due to internal loading) the fish community composition and biomass often change substantially. In Lake Constance a major reduction in total catches of fish and an increasing share of whitefish (*Coregonus lavaretus*) of the total commercial catch were observed (Box 2). For Danish lakes Jeppesen et al. (2002) found a significant reduction in the abundance of planktivorous fish and an increase in the share of potential piscivores at reduced TP, and in some lakes also in the abundance of potentially predatory fish not least perch (*Perca fluviatilis*) (Box 1), in some lakes after a delay of several years, however (Box 3). Likewise, studies in Swedish lakes also showed reduced abundance of cyprinids in gill net catches in shallow Lake Hjälmarn and of whitefish in the commercial catches in the more oligotrophic and deep Lake Vättern following the introduction of P- treatment at local sewage plants in the 1970s (Degerman et al. 2001).

Zooplankton are suggested to respond to both changes in available food and to predation. Yet, only few other studies have included the response of zooplankton to re-oligotrophication. In Lake Lugano, Switzerland, zooplankton biomass remained constant but eventually increased eight years after the TP concentration started to decline (Polli and Simona 1992), coinciding with a decrease in the commercial fish yield, particularly of cyprinids (Müller and Meng 1992). A stepwise change in zooplankton community structure and water clarity was also found in Lake Washington, USA (Edmondson and Lehman 1981), this being attributed to a decrease in the abundance of the invertebrate predator *Mysis relicta*. In contrast, Köhler et al. (2000) found a major decline in zooplankton biomass and the share of *Daphnia* in Lake Müggelsee during re-oligotrophication. For Danish lakes no significant changes in zooplankton biomass were found. Likewise, only minor changes were found in Lake Constance in response to a substantial TP reduction accompanied by a reduction in phytoplankton biomass (Straile and Geller 1998; Gaedke 1998). Release of fish predation counteracting the decline in food resources may be a likely explanation of the lack of changes in zooplankton biomass. Reduced predation on zooplankton is usually reflected by an increase in cladoceran mean size and in the share of *Daphnia* to total cladoceran abundance or biomass in the pelagial (Hrbácek et al. 1961; Brooks and Dodson 1965). In Danish lakes the share of *Daphnia* increased in several of the lakes studied, while their contribution in Lake Constance increased on an annual basis, but only negligibly so during summer.

In spite of improvements in some lakes, many lakes have exhibited delayed recovery or have not responded at all (Sas 1989). For some lakes this reflects insufficient reduction of nutrient input to trigger a shift to an ecologically more acceptable clear state. Thus, very significant and sustaining changes in the biological community and water transparency of shallow temperate freshwater lakes cannot be expected unless the TP concentration has been reduced to below 0.05-0.1 mg P l^{-1} (Jeppesen et al. 2000) and for deep lakes below 0.02 mg P l^{-1} (Sas 1989). For many lakes a slow response in lake water TP to reductions in external TP loading is attributed to internal loading, i.e. the P concentrations remain high because of TP release from the sediment pool accumulated when loading was high (Sas 1989; Søndergaard et al. 2001).

Box 3 Example of long-term resistance to external phosphorus loading reduction – Lake Søbygård, Denmark

Lake Søbygård is a small (0.39 km^2) shallow (mean depth 1.0 m, max depth 1.9 m) Danish lake. The hydraulic retention time is short (annual mean 18-27 days). Formerly, the lake received large amounts of only mechanically treated sewage water. In 1976, biological treatment was initiated at the treatment plant leading to a 3-10 fold reduction in loading of organic matter. In 1982, P-stripping was introduced at the plant and TP loading was markedly reduced. Despite the significant reduction in TP loading the ecological state improved only slowly (Fig. 2). The explanation is high internal P loading from the phosphorus pool accumulated in the high loading period. The net retention of phosphorus is still negative, i.e. more phosphorus leaves the lake than is received via the inlet. The sediment pool is only slowly reduced (Søndergaard et al. 2001) and predictions based on a simple mass balance model suggest that the lake in 2016 will reach a new steady state adapted to the present external loading, i.e. 34 years after the loading reduction (J.P. Jensen et al. unpublished data). The biological response has also been slow. The fish community remained dominated by cyprinids and the fish biomass (as judged from catches in multi-mesh-sized gillnets) remained high for many years (Fig. 2). Yet, 14 years after the loading reduction the percentage of piscivorous fish (mainly perch) increased substantially and the abundance of planktivorous fish and accordingly algal biomass expressed as chlorophyll *a* declined.

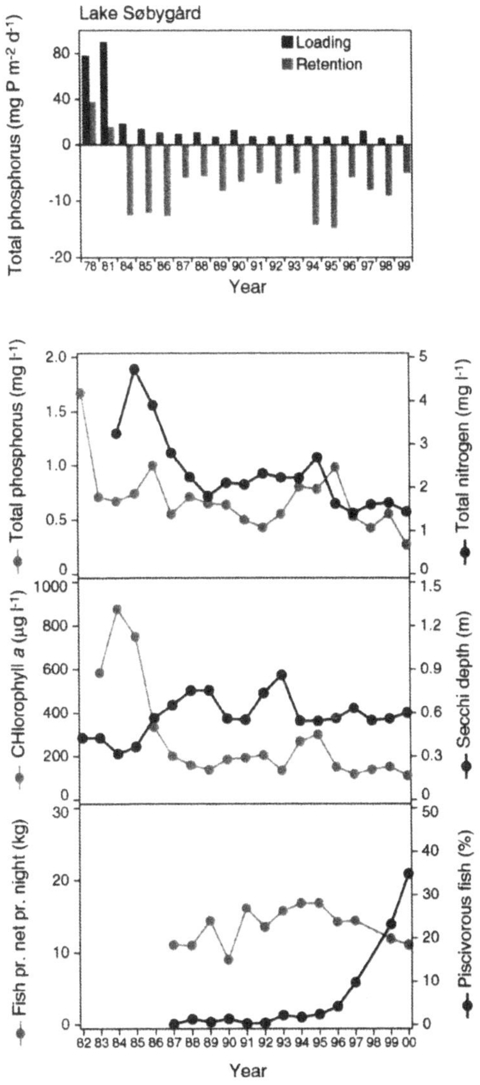

Fig. 2.

The development in Lake Søbygård is an illustrative example of significant chemical resistance owing to high internal P loading following a prolonged period with high external P loading.

Fig. 2. Top: Annual mean external loading and retention of phosphorus in Lake Søbygaard 1978-1999. Where retention is negative more phosphorus leaves than enters the lake. Below: Summer mean (May 1 – Oct. 1) concentrations of total phosphorus and nitrogen, algal biomass expressed as chlorophyll *a* and Secchi depth (lake water transparency) in the lake as well as the biomass of fish caught and percentage of piscivorous fish in multiple mesh-sized gill nets (42 m x 1.5 m, 14 different mesh sizes 6.25 – 75 mm). (Sources: Jeppesen et al. 1998, Søndergaard et al. 1999)

Many years may pass before the surplus pool is released or permanently buried. The duration of this transitional period depends on, for instance, the duration of the pollution period, the residence time and the supply of P binding agents from the surroundings, such as iron and calcium. A fast reduction of TP levels has been observed in some lakes (Box 2), while delays of as much as 20-40 years have been observed in others (Box 3).

Internal loading is also affected by biological homeostasis. Planktivorous and benthivorous fish in particular contribute to biological resistance in lakes (Jeppesen and Sammalkorpi 2002). A continuously high fish predation prevents both the appearance of large herbivorous zooplankton that would otherwise clear the water and diminishes the number of benthic animals stabilizing and oxidizing the sediment. Moreover, excretion of nutrients to overlaying waters from benthic-feeding fish or bioturbation by fish foraging amongst the surface sediment at increasing turbidity may play a role (Andersson et al. 1978; Brabrand et al. 1990; Breukelaar et al. 1994; Horppila et al. 1998). Extinction of light prevents the growth of benthic algae and the appearance of submerged macrophytes, thus maintaining low sediment retention capacity.

Based on an analysis of 18 case studies from European lakes, Sas (1989) discussed the response of lakes being in the transient phase until a new equilibrium state adapted to the new and lower nutrient loading is reached. He proposed the following general model for the response of lake water total phosphorus concentrations:

$$TP_{lake}\,post = TP_{lake}\,pre\,(TP_{inlet}\,post/\,TP_{inlet}\,pre)^{0.65} \quad (1)$$

where:
TP_{lake} = in-lake total P concentration (May-Oct.); TP_{inlet} = discharge weighted annual mean inlet total P concentration (annual), pre = pre-treatment period; post = expected in new equilibrium state. The exponent is lower than predicted in the OECD steady state model (0.82), which was attributed to internal loading.

The chlorophyll *a* level (Chla) expected at equilibrium was modelled analogously to equation 1:

$$Chla_{lake}\,post = Chla_{lake}\,pre\,(P_{lake}\,post/\,P_{lake}\,pre)^{k} \quad (2)$$

where $Chla_{lake}$ = summer chlorophyll *a* concentration (May-Oct.) $\mu g\ l^{-1}$; P_{lake} = total P concentration in the lake (annual), pre = pre-treatment period; post = new equilibrium state, k is 0.6 for deep lakes and 1.4 for shallow lakes. These relationships must be used with caution, as they are based on short time series (typically 5-10 years), as also argued by Sas (1989). Others have developed general mass balance models including simple or more complex models for internal loading (see Cooke et al. 1993).

Lake Restoration in Northern Temperate Lakes

Various methods have been used to reduce the internal loading of phosphorus, including sediment removal and chemical treatment of the sediment with aluminum or iron salts. In stratified lakes also injections of oxygen or nitrate to the hypolimnion or destabilization of the thermocline have been used (Cooke et al. 1993).

To combat biological resistance, a number of fish manipulation methods have been developed. A widely used method in the temperate zone is selective removal of planktivorous fish in order to enhance grazer control of phytoplankton by releasing predation on zooplankton. Removal of 75-80% of the stock of planktivorous and benthivorous fish during 1-2 years is recommended in order to avoid their re-growth and to stimulate the growth of potentially piscivorous perch (Perrow et al. 1997; Meijer et al. 1999; Hansson et al. 1998; Jeppesen and Sammalkorpi 2002). An alternative or supplementary method to fish removal is the stocking of high numbers of 0^+ pike (*Esox lucius*) (>1000 ha^{-1}) to control newly hatched plankti-benthivorous fish such as roach (*Rutilus rutilus*) and bream (*Abramis brama*) (Prejs et al. 1994; Berg et al. 1997), though the results obtained are ambiguous (Skov and Berg 1999). Also stocking of pikeperch (*Stizostedion lucioperca*) has been used (Benndorf 1995).

The results so far indicate that fish manipulation may have a long-term effect in shallow temperate lakes if nutrient loading is reduced to a TP level below 0.05-0.1 mg P l^{-1} in the future state of equilibrium (Jeppesen and Sammalkorpi 2002), although the long-term stability of such a measure is still subject to debate. If nitrogen loading is low, stocking may have a positive impact at higher phosphorus concentrations. The threshold of 0.05-0.1 mg P l^{-1} is in accordance with the major changes in lake biological structure that usually occur within this range. Yet, temporary effects of biomanipulation can be obtained in lakes with high nutrient concentrations, but it is unlikely that the effect will prevail in such lakes in the long term unless periodical reduction in the abundance of planktivorous fish is conducted.

Fish manipulation may affect the nutrient level as well. A 30-50% reduction in lake concentrations of total N and total P has been found in the most successful fish manipulation experiments, even when macrophytes were absent (Jeppesen et al. 1998; Søndergaard et al. 2001). This is attributed to increased growth of microbenthic algae that trap nutrients, which stimulates fixation of inorganic N and P and the denitrification in the sediment. Moreover, microbenthic algae and reduced sedimentation of phytoplankton improve redox conditions, thereby reducing phosphorus release from the sediment. Further, elevated abundance of benthic invertebrates following release from fish predation may stabilize the sediment, thereby reducing the risk of resuspension. Finally, decreased release of nutrients to overlaying waters from benthic-feeding fish may contribute importantly to a reduction.

Grazing by herbivorous waterfowl like coot (*Fulica atra*) and mute swan (*Cygnus olor*) may also create resistance by delaying submerged macrophyte recolonization. (Mitchell and Perrow 1997; Perrow et al. 1997). Establishment of enclosures to protect macrophytes against grazing by waterfowl and herbivorous fish has also been employed as an alternative or supplementary restoration tool (Lauridsen et al. 1993; Søndergaard et al. 1996). The exclosures enable the macrophytes to grow in a grazer-free environment from where they may spread seeds, turions or plant fragments augmenting colonization. Moreover, they can be used as a valuable daytime refuge for zooplankton. The usefulness of plant refuges as a restoration tool is probably limited to the same nutrient level as the fish manipulation methods, and their effect is most likely particularly high in small lakes where the density of herbivorous waterfowl is highest in the absence of plants (Jeppesen et al. 1999).

Lake Restoration in Southern Temperate, Subtropical and Tropical Lakes

Experience with lake restoration in south temperate, subtropical and tropical lakes is far less extensive than with northern temperate lakes. However, some studies have shown that a nutrient loading reduction may lead to improvement of the ecological state via declining algal biomass and increased lake water transparency (Lake Apopka, Florida) (Lowe et al. 2001). For multiple reasons it is debatable whether the fish manipulation methods described above for northern temperate lakes are usable as a supplementary tool in order to speed up the recovery process in subtropical and tropical lakes (Lazarro 1997). First, omnivory is far more common in warm water lakes, where several species prey on both zooplankton and phytoplankton. Second, the share of piscivorous fish that may potentially control the planktivorous fish is often low and, third, due to the high temperatures the spawning period of planktivorous fish is prolonged and several species have more than one cohort per year (Fernando 1994), which further augments the predation pressure on the zooplankton. It is thus characteristic that warm water lakes are generally dominated by small zooplankton species (Dumont 1994). It is therefore likely that a reduction in the biomass of planktivores by fish removal will be compensated by fast re-growth in the remaining population and the impact will therefore be of only short duration. It should, however, be emphasized that the current basis of experience within this field is inadequate (Lazarro 1997), as it is as well for southern temperate lakes (Howard-Williams and Kelly this volume).

Future Perspectives

Within the past two decades major advances have occurred in our knowledge of trophic interactions in northern temperate lakes and of the measures to be applied for lake restoration purposes when a shift from the clear to the turbid state has occurred. To restore the lakes, the main triggering factor has to be controlled, namely the input of nutrients to the lake. In the subsequent phase both fast responses and significant resistance have been observed, the resistance being attributed to internal loading and biological homeostasis. A number of general conceptual and empirical models have been developed describing this transient phase after loading reduction until a new equilibrium state is reached, as well as the biological interactions among biota and internal loading. Significant improvements of these models are certain to occur in future concurrently with an increasing number of case-studies and longer time series. However, exceptions will always exist. As described nicely by Howard-Williams and Kelly (this volume), the general models serve as a good starting point, but one should always be aware that local conditions may be of essential importance. In some lakes, for example, nitrogen may be the limiting nutrient for phytoplankton growth and the general phosphorus-based models are thus not applicable here (Moss 1998). General models should therefore be applied with caution, and if used on their own can lead to ecologically as well as economically unwise decisions.

Various additional restoration measures can be initiated of which especially biomanipulation is attractive as it is typically much less expensive than the traditionally applied physico-chemical methods. However, care must be taken not to introduce exotic species to obtain a rapid response, as this may lead to serious problems later on. When choosing restoration methods local factors must always be considered.

While the response of northern temperate lakes to loading reductions has been well elucidated, the lack of general knowledge of and models for south temperate, subtropical and tropical lakes is striking. Procurement of such information through experimental studies is of vital importance as the eutrophication problems are expected to increase dramatically in subtropical and tropical lakes in the future.

Acknowledgements

We thank the Danish counties for access to data from the National Monitoring Programme of Danish Lakes. The study was supported by the Danish Natural Science Research Council (grant 9601711) and the research programme "The role of fish in ecosystems, 1999-2001" funded by the Ministry of Agriculture, Fisheries and Food, ECOFRAME and BIOMAN funded by EU, and the project "Consequences of weather and climate changes for marine and freshwater ecosystems: Conceptual and operational forecasting of the aquatic environment", supported by the Danish Research Council. We also thank Juana Jacobsen, Kathe Møgelvang and Anne Mette Poulsen for technical assistance. Furthermore, we wish to thank Michio Kumagai for inviting EJ to participate in the international workshop in Otsu, Japan, 2001 and for encouraging us to write this paper. Finally, we are grateful to Clive Howard-Williams for discussions at the meeting and for valuable critical comments on our paper.

References

ANDERSSON, A., H. BERGGREN, AND G. CRONBERG. 1978. Effects of planktivorous and benthivorous fish on organisms and water chemistry in eutrophic lakes. Hydrobiologia **59:** 9-15.

ANNEVILLE, O., AND J. P. PELLETIER. 2000. Recovery of Lake Geneva from eutrophication: quantitative response of phytoplankton. Arch. Hydrobiol. **148:** 607-624.

BEKLIOGLU, M., L. CARVALHO, AND B. MOSS. 1999. Rapid recovery of a shallow hypertrophic lake following sewage effluent diversion: lack of chemical resilience. Hydrobiologia **412:** 5-15.

BENNDORF, J. 1995. Possibilities and limits for controlling eutrophication by biomanipulation. Int. Rev. Ges. Hydrobiol. **80:** 519-534.

BERG, S., E. JEPPESEN, AND M. SØNDERGAARD. 1997. Pike (*Esox lucius* L.) stocking as a biomanipulation tool. 1: Effects on the fish population in Lake Lyng (Denmark). Hydrobiologia **342/343:** 311-318.

BERNHARDT, VON H., J. CLASEN, O. HOYER, AND W. WILHELMS. 1985. Oligotrophication in lakes by means of chemical nutrient removal from the tributaries. Its demonstration with the Wahnbach Reservoir. Arch. Hydrobiol. Suppl. **70:** 481-533.

BRABRAND, Å., B. FAAFENG, AND J. P. NILSSEN. 1990. Relative importance of phosphorus supply to phytoplankton production: fish excretion versus external loading. Can. J. Fish. Aquat. Sci. **47:** 364-372.

BREUKELAAR, A.W., E. H. R. R. LAMMENS, J. P. G. KLEIN BRETELER, AND I. TATRAI. 1994. Effects of benthivorous bream (*Abramis brama* L.) and carp (*Cyprinus caprio* L.) on sediment resuspension and concentration of nutrients and chlorophyll *a*. Freshwat. Biol. **32:** 113-121.

BROOKS, J. L., AND S. I. DODSON. 1965. Predation, body size and composition of plankton. Science **150:** 28-35.

BÄUERLE, E., AND U. GAEDKE [eds.], 1998. Advances in Limnology 53. Lake Constance: Characterization of an ecosytem in transition.
E. Schweizerbart'sche Verlagsbuchhandlung, 610p.

CHOULIK, O., AND T. R. MOORE. 1992. Response of a subarctic lake chain to reduced sewage loading. Can. J. Fish. Aquat. Sci. **49:** 1236-1245.

COOKE, G. D., E. B. WELCH, S. A. PETERSON, AND P. R. NEWROTH. 1993. Restoration and management of lakes and reservoirs, Boca Raton, Florida: Lewis Publishers.

CRONBERG, G. 1999. Qualitative and quantitative investigations of phytoplankton in Lake Ringsjön, Scania, Sweden. Hydrobiologia **404:** 27-40.

DEGERMAN, E., J. HAMMAR, P. NYBERG, AND G. SVÄRDSON. 2001. Human impact on the fish density in four largest lakes of Sweden. Ambio. **30:** 522-528.

DUMONT, H. J. 1994. On the diveristy of the Cladocera in the tropics. Hydrobiologia **272:** 27-38.

ECKMANN, R., AND R. RÖSCH. 1998. Lake Constance fisheries and fish ecology. Arch. Hydrobiol. Spec. Issues Advanc. Limnol. **53:** 285-301.

EDMONDSON, W. T., AND J. T. LEHMAN. 1981. The effect of changes in the nutrient income on the condition of Lake Washington. Limnol. Oceanogr. **26:** 1-29.

FERNANDO, C. H. 1994. Zooplankton, fish and fisheries in tropical freshwaters. Hydrobiologia **272:** 105-123.

GAEDKE, U. 1998. The response of the pelagic food web to re-oligotrophication of a large and deep lake (L. Constance): Evidence for scale-dependent hierarchical patterns? Arch. Hydrobiol. Spec. Issues Advanc. Limnol. **53:** 317-333.

GÜDE, H., H. ROSSKNECHT, AND G. WAGNER. 1998. Anthropogenic impacts on the trophic state of Lake Constance during the 20th century. Arch. Hydrobiol. Spec. Issues Advanc. Limnol. **53:** 85-108.

HANSON, J. M., AND W. C. LEGGETT. 1982. Empirical prediction of fish biomass and weight. Can. J. Fish. Aquat. Sci. **39:** 257-263.

_________, AND R. H. PETERS. 1984. Empirical prediction of crustacean zooplankton biomass and profundal macrobenthos biomass in lakes. Can. J. Fish. Aquat. Sci. **41:** 439-445.

HANSSON, L-A., H. ANNADOTTER, E. BERGMAN, S. F. HAMRIN, E. JEPPESEN, T. KAIRESALO, E. LUOKKANEN, P-Å. NILSSON, M. SØNDERGAARD, AND J. STRAND. 1998.

Biomanipulation as an application of food chain theory: constraints, synthesis and recommendations for temperate lakes. Ecosystems **1:** 558-574.

HORPPILA, J., H. PELTONEN, T. MALINEN, E. LUOKKANEN, AND T. KAIRESALO. 1998. Top-down or bottom-up effects by fish – issues of concern in biomanipulation op lakes. Restor. Ecol. **6:** 1-10.

HOWARD-WILLIAMS, C., AND D. KELLY. 2003. Recovery from eutrophication: Local perspectives in lake restoration and rehabilitation, Chapter 5-2. *In* M. Kumagai and W. F. Vincent [eds.], Freshwater Management: Global Versus Local Perspectives. Springer-Verlag, Tokyo, this volume.

HRBÁCEK, J., V. DVORAKOVA, V. KORINEKAND, AND L. PROCHAZKOVA. 1961. Demonstration of the effect of the fish stock on the species composition of zooplankton and the intensity of metabolism of the whole plankton association. Verh. Internat. Verein. Limnol. **14:** 192-195.

JEPPESEN, E., J. P. JENSEN, M. SØNDERGAARD, T. LAURIDSEN, L. J. PEDERSEN, AND L. JENSEN. 1997. Top-down control in freshwater lakes: the role of nutrient state, submerged macrophytes and water depth. Hydrobiologia **342/343:** 151-164.

________, ________, ________, ________, P. H. MØLLER, AND K. SANDBY. 1998. Changes in nitrogen retention in shallow eutrophic lakes following a decline in density of cyprinids. Arch. Hydrobiol. **142:** 129-151.

________, M. SØNDERGAARD, J. P. JENSEN, E. MORTENSEN, A.-M. HANSEN AND T. JØRGENSEN. 1998. Cascading trophic interactions from fish to bacteria and nutrients after reduced sewage loading: an 18-year study of a shallow eutrophic lake. Ecosystems **1:** 250-267.

________, J. P. JENSEN, M. SØNDERGAARD, AND T. LAURIDSEN. 1999. Trophic dynamics in turbid and clearwater lakes with special emphasis on the role of zooplankton for water clarity. Hydrobiologia **408/409:** 217-231.

________, ________, ________, ________, AND F. LANDKILDEHUS. 2000. Trophic structure, species richness and biodiversity in Danish lakes: changes along a phosphorus gradient. Freshwat. Biol. **45:** 201-213.

________, ________, AND ________. 2002. Response of phytoplankton, zooplankton and fish to re-oligotrophication: an 11-year study of 23 Danish lakes. Aquat. Ecosys. Health & Managm. **5:** 31-43.

________, AND I. SAMMALKORPI, 2002. Lakes, Chapter 14, p. 297-324. *In* M. Perrow and T. Davy [eds.], Handbook of Restoration Ecology. Cambridge University Press.

KÖHLER, J., H. BERNHARDT, AND S. HOEG. 2000. Long-term response of phytoplankton to reduced nutrient load in the flusehd Lake Müggelsee (Spree system, Germany). Arch. Hydrobiol. **148:** 209-229.

KÜMMERLIN, R. E. 1998. Taxonomical response of the phytoplankton community of Upper Lake Constance (Bodensee-Obersee) to eutrophication and re-oligotrophication. Arch. Hydrobiol. Spec. Issues Advanc. Limnol. **53:** 109-117.

LATHROP, R. C. 1990. Response of Lake Mendota (Wisconsin, U.S.A.) to decreased phosphorus loadings and the effect on downstream lakes. Ver. Int. Verein. Limnol. **24:** 457-463.

LAURIDSEN, T., E. JEPPESEN, AND F.Ø. ANDERSEN. 1993. Colonization of submerged macrophytes in shallow fish manipulated Lake Væng: Impact of sediment composition and birds grazing. Aquat. Bot. **46:** 1-15.

LAZARRO, X. 1997. Do the trophic cascade hypothesis and classical biomanipulation approaches apply to tropical lakes and reservoirs? Ver. Int. Verein. Limnol. **26:** 719-730.

LOWE, E. F., L. E. BATTOE, M. F. COVENY, C. L. SCHELSKE, K. E. HAVENS, E. R. MARZOLF, AND K. R. REDDY. 2001. The restoration of Lake Apopka in relation to alternative stable states: an alternative view to that of Bachmann et al. (1999). Hydrobiologia **448:** 11-18.

MARSDEN, S. 1989. Lake restoration by reducing external phosphorus loading: the influence of sediment phosphorus release. Freshw. Biol. **21:** 139-162.

MEIJER, M.-L., I. DE BOOIS, M. SCHEFFER, R. PORTIELJE, AND H. HOSPER. 1999. Biomanipulation in shallow lakes in the Netherlands: an evaluation of 18 case studies. Hydrobiologia **408/409:** 13-30.

MITCHELL, S. F., AND M. R. PERROW. 1997. Interactions between grazing birds and macrophytes, p. 175-196. *In* E. Jeppesen, Ma. Søndergaard, Mo. Søndergaard and K. Christoffersen [eds.], The structuring role of submerged macrophytes in lakes. Ecological Studies, Vol. 131. Springer Verlag, New York.

MOSS, B. 1998. Shallow lakes: biomanipulation and eutrophication. Scope Newsletter 29, 45p.

MÜLLER, R., AND H. J. MENG. 1992. Past and present state of the ichthyofauna of Lake Lugano. Aquat. Sci. **54:** 338-350.

OECD. 1982. Eutrophication of waters. Monitoring, assessments and control. OECD, Paris. 210 p.

OLSÉN, P., AND E. WILLÉN. 1980. Phytoplankton response to sewage reduction in Vättern, a large oligotrophic lake in Central Sweden. Arch. Hydrobiol. **89:** 171-188.

PERROW, M. P., M-L. MEIJER, P. DAWIDOWICZ, AND H. COOPS. 1997. Biomanipulation in shallow lakes: state of the art. Hydrobiologia **342/343:** 355-363.

POLLI, B., AND M. SIMONA. 1992. Qualitative and quantitative aspects of the evolution of the planktonic populations in Lake Lugano. Aqat. Sci. **54:** 303-320.

PREJS, A., A. MARTYNIAK, S. BORON, P. HLIWA, AND P. KOPERSKI. 1994. Food web manipulation in small, eutrophic Lake Wirbel, Poland: effect of stocking with juvenile pike on planktivorous fish. Hydrobiologia **275/276:** 65-70.

REYNOLDS, C. F. 1984. The ecology of freshwater phytoplankton. Cambridge University Press. Cambridge, 384p.

RUGGIO, D., G. MORABITO, P. PANZANI, AND A. PUGNETTI. 1998. Trends and relations among basic phytoplankton characteristics in the course of the long-term oligotrophication of Lake Maggiore (Italy). Hydrobiologia **370:** 243-257.

SAS, H. [ed.], 1989. Lake restoration by reduction of nutrient loading. Expectation, experiences, extrapolation. Acad. Ver. Richardz Gmbh.

SKOV, C., AND S. BERG. 1999. Utilisation of natural and artificial habitats by YOY pike in a biomanipulated lake. Hydrobiologia **408/409:** 115-122.

STRAILE, D., AND W. GELLER. 1998. Crustacean zooplankton in Lake Constance from 1920 to 1995: Response to eutrophication and re-oligotrophication. Arch. Hydrobiol. Spec. Issues Advanc. Limnol. **53:** 255-274.

SØNDERGAARD, M., L. OLUFSEN, T. LAURIDSEN, E. JEPPESEN, AND T. V. MADSEN. 1996. The impact of grazing waterfowl on submerged macrophytes: in situ experiments in a shallow eutrophic lake. Aquat. Bot. **53:** 73-84.

_________, J. P. JENSEN, AND E. JEPPESEN. 1999. Internal phosphorus loading in shallow Danish lakes. Hydrobiologia **408/409:** 145-152.

_________, _________, AND _________. 2001. Retention and internal loading of phosphorus in shallow, eutrophic lakes. The Scientific World **1:** 427-442.

_________, E. JEPPESEN, AND P. H. MØLLER. 2002. Seasonal dynamics in the concentrations and retention of phosphorus in shallow Danish lakes during recovery. Aquat. Health Managm. **5:** 19-29.

UN. 1998. World population prospects: The 1996 revision. United Nations Secretariat, Department of Economic and Social Affairs, Population Division. Volume 1, 614 pp.

VOLLENWEIDER, R. A. 1976. Advance in defining critical loading levels for phosphorus in lake eutrophication. Mem. Ist. Ital. Idrobiol. **33:** 53-83.

WETZEL, R. G. 1990. Land-water interfaces: metabolic and limnological regulators. Verh. Int. Verein. Limnol. **24:** 6-24.

WILLÉN, E. 1984. The large lakes of Sweden, Vänern, Vättern, Mälaren and Hjälmaren, p. 107-134. *In* F. B. Taub [ed.], Ecosystems of the World 23. Lakes and Reservoirs. Elsevier.

________. 2001. Four decades of research on the Swedish large Lakes Mälaren, Hjälmaran, Vättern and Värnern: the significance of monitoring and remedial measures for a sustainable society. Ambio **30:** 458-466.

WOJCIECHOWSKI, I., W. WOJCIECHOWSKA, K. CZERNAS, J. GALEK, AND K. RELIGA. 1988. Changes in phytoplankton over a ten-year period in a lake undergoing de-eutrophication due to surrounding peat bogs. Arch. Hydrobiol. Suppl. **78:** 373-387.

5-2. Local Perspectives in Lake Restoration and Rehabilitation

Clive Howard-Williams and David Kelly

National Institute of Water and Atmospheric Research Ltd,
P.O. Box 8602, Christchurch, New Zealand

Abstract

Eutrophication processes resulted in a serious deterioration of lakes across the world between the 1950s and the late 1970s. Over the last 20 years there have been efforts in many countries to halt the process of eutrophication, and to restore lakes to some "ideal" state. There have been three phases in this process: Reduction or removal of point source discharges of nutrients; Reduction of diffuse source discharges of nutrients; Physical or biological manipulation of lakes. There have been attempts to combine the knowledge from these individual cases to produce general predictive models of the process of lake restoration following nutrient reduction. There are three major factors that tend to reduce the applicability of general predictive models of lake restoration, and they clearly operate at local scales. These are: (1) the local geological context, which among other things determines the limiting nutrient (eg nitrogen vs phosphorus) and the extent to which groundwater affects nutrient inputs to lakes from the catchment, (2) the effect of climate (which influences the stratification and the extent of internal loading), and (3) regional or local biogeography. Examples are provided for lakes that do not conform to traditional Northern Hemisphere eutrophication models and where restoration needs to be specific to local conditions. These are lakes of the volcanic areas that are nitrogen limited due to local P-rich soils, and lakes of some south temperate latitudes that have the peak phytoplankton biomass in winter due to local or regional climatic conditions influencing the stratification cycle. The importance of local biogeography and biodiversity is a key factor on restoration processes. Some areas simply have no herbivorous fish for instance, and in other areas invasive species have so modified lake systems that the goals of restoration are inherently compromised. We argue that local perspectives are likely to override conclusions derived from global models of the lake restoration process. As a consequence, a recommended approach is to use some type of expert system tool applied to lakes on a case-by-case basis. We also recognise that linking lakes with the people of the lake catchment is central to any programme. It is local communities who decide on the goals of the restoration taking into account affordability and local cultural ideals.

Introduction

Eutrophication processes resulted in a serious deterioration of lakes across the world between the 1950s and the late 1970s. In some countries this trend continues e.g. China (Jin 1994). The classical studies on the eutrophication process were concerned with the relationships of the summer chlorophyll a concentration in lake waters as a function of the phosphorus concentration (Sakamoto 1966, Dillon and Rigler 1974). Vollenweider (1979) combined these correlations with a nutrient loading model to relate spring-time chlorophyll biomass to nutrient load. As models developed to assist with the remediation of lake eutrophication, the corollary of the enrichment relationships was that if phosphorus concentration was reduced, chlorophyll reduced according to the same relationships. While the enrichment models were useful in a broad sense, the relationships were usually in a log-log form. Therefore, for individual lakes, the errors on the predictions were very large and it was not surprising that reductions in phosphorus loads were often not matched by corresponding reductions in algal biomass. Over the last 20 years there have been efforts in many countries to halt the process of eutrophication, and to restore lakes to some state, ideally to their pre-eutrophic condition. This chapter argues that the influences of local physical, chemical and biological processes are sufficiently large to require a case-by-case approach to restoration. In addition, the restoration of lakes closely involves people's perceptions of what a healthy lake ecosystem should be.

There have been three historical phases in the process of lake restoration:

1. Reduction or removal of **point source** discharges of nutrients. This phase has been relatively easy to initiate in many countries by simple legislation requiring discharge permits that link to national environmental standards.

2. Reduction of **diffuse source** discharges of nutrients. This has been, and is, a more difficult process to initiate and enforce. In New Zealand and Australia for instance, population densities in lake catchments are generally very much lower than typically found in Europe and North America (Vant 1987a), although a high diffuse source nutrient supply is well documented for arable areas of Europe. The Grebiner See in North Germany, as one example, has 95% of its nutrient input from diffuse sources (Ohle 1982). Eutrophication problems in New Zealand generally reflect the increase in diffuse nutrient loads associated with the intensification of agriculture (White 1982). Nutrient run-off from land directly into streams or to groundwater is difficult to control and regulate. It can be done only by catchment management techniques, soil conservation measures and specific management practises such as wetland and riparian conservation, rehabilitation or planting (Vant 1987 b, Carpenter and Lathrop 1999, Jeppesen et al. 1999).

3. Physical or biological **manipulation** of lakes. This involves re-establishing species or communities lost by degradation, changing the food-web by, for instance, deliberate introductions of species (Jeppesen et al. 1999), removing invasive species, and /or by combining these methods with physical techniques such as hypolimnetic aeration, hypolimnetic nitrate treatment etc. (Vant 1987b, Harper et al. 1999, Sondergaard et al. 2000).

For a long-term sustainable restoration programme the first two and sometimes all three of the above are needed, in combination with a social fabric that is prepared to meet the cost of a restoration programme. Often the goal for the restoration programme has not been clearly set prior to expenditure on a remediation or restoration programme. This neglect, or at least inadequate definition, has frequently been responsible for subsequent shortcomings of the restoration programme (Moss et al. 2002). Considerable advances in useful technologies for lake restoration have been made, and yet, as pointed out by Carpenter and co-workers (Carpenter et al. 1998, Carpenter and Cottingham 1997, Carpenter and Lathrop 1999) lake restoration is difficult achieve on a wide spread basis. The literature suggests that the most useful models and concepts have been derived for shallow lakes, rather than deep lakes, and the bulk of the work reported in the international literature appears to have been concentrated on shallow lakes. Many of these are small, and clearly logistically easier to subject to a restoration programme. Some countries (e.g., the Netherlands) have tried lake restoration at a national level (Portielje and Van der Molen 1999), but in most cases restoration has been at a local scale. Attempts have been made to combine the knowledge from individual cases to produce general predictive models of the process of lake restoration following nutrient reduction (Phillips et al. 1999). However the success of lake restoration programs still varies to a large degree that, we argue, is influenced by processes operating at localized scales.

Factors Influencing the Applicability of General Models

There are three interlinking factors that tend to reduce the applicability of general predictive models of lake restoration (such as the summer chlorophyll vs phosphorus loading models). These factors clearly operate at local scales, with the result that the restoration process of each lake must be considered on a case-by-case basis. The three factors are: (1) the local geological context which, among other things, determines the limiting nutrient and the extent to which groundwater affects nutrient inputs to lakes from the catchment; (2) the effect of climate which influences both lake mixing and stratification cycle and the extent of internal nutrient loading; (3) regional or local biogeography and biodiversity. Although most of the examples we cite are for lakes contained within New Zealand, many of these concepts also pertain to other portions of the globe, particularly temperate lakes in the southern hemisphere (e.g., southern Australia, Argentina, Chile).

The Geological Context — The first point we wish to consider in developing lake restoration models is the importance of considering local geological effects on physio-chemical processes of the lake, and how this interacts with lake restoration measures taken within the catchment (e.g., diffuse source nutrient control). Local geology influences soil type, landform and lake morphometry, and it is the main determinant of water chemistry in the particular waterbody. It also determines the interaction of groundwater and surface water and in particular it influences the nutrient status of lakes. Under this heading there are two particular factors of importance. These are: the limiting nutrient and the influence of groundwater.

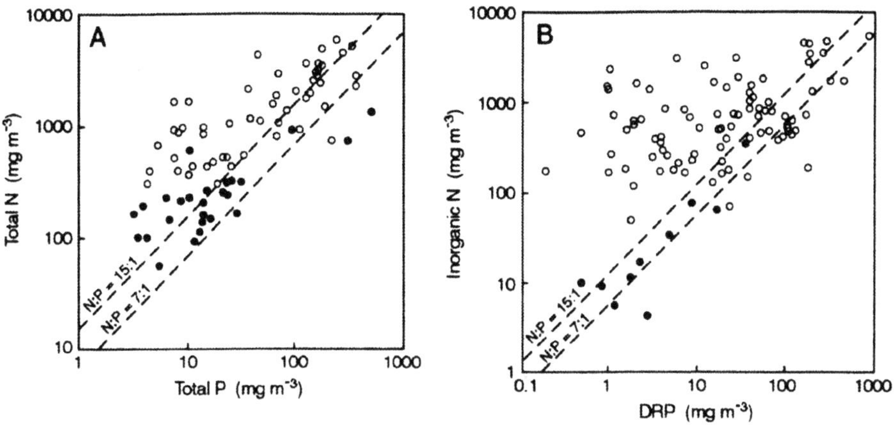

Fig. 1. Relative distribution of average annual values for nutrients in the, largely northern temperate, lakes of the OECD study (○) and New Zealand lakes (●). A. Total nitrogen and total phosphorus concentrations. B. Inorganic nitrogen and dissolved reactive phosphorus concentrations. The lines define a zone of N:P ratios that are typical of balanced algal growth (OECD 1982). From Viner and White (1987).

The Limiting Nutrient — Much of the science behind eutrophication and subsequent restoration has been derived from northern hemisphere industrial or agriculturally rich countries where phosphorus is the limiting nutrient. This led to comments, such as those in Golterman (1975), that "The inescapable conclusion therefore is that there is only one remedy for the excessive growth of algae: diminished input of phosphorus into lakes by removing them from sewage effluents and by replacing polyphosphates in detergents with a phosphate-free product." However, in many areas defined by local geology, this emphasis on phosphorus needs to be questioned (Goldman and Horne 1983, White 1983, Burns 1991). Volcanic soils are generally rich in phosphorus and often poor in nitrogen and for lakes in volcanic catchments, nitrogen is frequently the limiting nutrient.

The lakes of the central volcanic plateau in New Zealand provide a well-documented example of a local geological effect, although a similar analysis might be just as appropriate for the volcanic catchments of Japan and East Africa. The average total nitrogen content of lake waters in the volcanic areas of New Zealand is less than half that of the average concentration in lakes of the OECD countries (Fig. 1, OECD 1982, White 1983). In terms of inorganic nitrogen in lake waters, the differences are even more striking. Where average inorganic P concentrations were less than 2 mg m^{-3}, the lakes of the OECD generally had total N concentrations in excess of 100 mg m^{-3}, while this value was less than 20 in the New Zealand volcanic plateau lakes and often, where measurable inorganic P is encountered, inorganic N is undetectable. Nitrogen limitation of phytoplankton growth has been recorded in volcanic plateau lakes in many instances. For instance, White et al. (1986) ran experiments on nutrient additions to 10 lakes from the North Island of New Zealand. These included chlorophyll responses to nutrient additions and a suite of short-term physiological assays and measured nutrient concentrations. In samples from all ten lakes phytoplankton biomass increase was prevented by a shortage of nutrient nitrogen, although the extent of this varied on a seasonal basis.

An excellent example of the non-critical use of a "global" approach to solving a local problem was seen in Lake Rotorua. In the 1970s eutrophication problems in this lake were exhibited in the form of surface scums comprising bloom-forming cyanobacteria. On the basis of north temperate lake experience, phosphorus stripping of the Rotorua City sewage discharge was recommended by European consultants as a means of lake rehabilitation. A phosphorus stripping plant was installed, but by the late 1980s it was considered to have been unsuccessful (Viner and White 1987). The reasons were twofold. Firstly, natural loads of phosphorus from the volcanic catchment exceeded the sewage-derived phosphorus loads into the lake, a situation apparently unknown in Northern Europe, and secondly, the phytoplankton of Lake Rotorua was limited by nitrogen rather than phosphorus. Subsequent nitrogen stripping in the late 1980s and early 1990s by land treatment of the effluent has had an effect, although internal loading is still seen as a short –term problem. In the long term it is recognised that the lake water quality is likely to improve by a combination of both nitrogen and phosphorus removal.

Groundwater — The degree to which groundwater provides a nutrient source influences the capacity for nutrient reduction. Clearly nutrient removal or biological assimilation is more easily carried out in surface waters than in groundwater. The importance of groundwater in lake hydrological cycles varies on a regional or local scale and over highly varying time scales. The residence time of most aquifers is long. For instance, nitrogen in groundwater entering Lake Taupo in New Zealand's North Island is believed to originate from farming practises 20 years ago (Vant and Huser 2000) while in the same lake, reticulation of sewage from lakeside settlements resulted in rapid (3-18 months) changes to groundwater nutrients entering the lake (Hawes and Smith 1993, John et al. 1978). There is little that can be done to remedy excess nutrients once they are in the groundwater and lake restoration plans need to consider the local, catchment by catchment, effects of groundwater contamination over appropriate timescales.

The degree to which a catchment is influenced by land use, relative wetland area and riparian protection of inflowing streams has a major impact on the efficiency of diffuse contaminant load reduction (Carpenter and Cottingham 1997) and hence on the rehabilitation potential for a lake. Positioning of the lake within the landscape and its interaction with surface and groundwater flow-paths strongly affects the susceptibility of lakes to eutrophication (Krantz et al. 1997, Devito et al. 2000). Lakes located in areas of groundwater recharge or local area discharges are more likely to experience greater inputs of nutrients from land-uses (e.g., agriculture or deforestation). Groundwater generally enters lakes at a depth of 1-2 m (unless the aquifer is confined by local geology) and plant communities of the littoral zone of lakes may attenuate much of the nutrient. Submerged macrophyte beds, particularly those with tall canopy forming species (eg *Potamogeton pectinatus*) have been shown to efficiently remove loads of up to 100 mg m^{-3} week $^{-1}$ of DRP and 1000 mg m^{-3} wk^{-1} of dissolved inorganic N (Howard-Williams 1986). Such loads are relatively low, but in the range of most ground-waters. Lake-edge wetlands have been shown to reduce nitrogen concentrations in groundwater moving through them by up to 98% (Gibbs and Matheson 2001). Because these properties are dictated by highly localized landscape effects, restoration models will need to consider these attributes in their design. Nutrient attenuation and transformations in shallow groundwater (top 1 –2m) can be strongly influenced by the degree of riparian protection (Downes et al. 1997). This is also a highly localised issue, dependent on landscape, agricultural use, river-edge land values and once more, the willingness of the local community (or the legislators) to protect land and stream margins. There can be no "global" model for this other than perhaps the general concept of protection of streams entering lakes. In New Zealand, for example, there are published guidelines for local governing authorities on the establishment and management of riparian strips along waterways (Collier et al. 1995).

The Influence of Climate — Two factors will be discussed under this heading: winter biomass and internal loading.

Winter Biomass — Land masses which are subject to maritime climatic influences (New Zealand, southern Chile and Argentina may be examples) tend to have lakes with a deep mixed layer encouraging the transfer of nutrients from hypolimnia to the surface over extended periods, from late summer to winter mixing. This is a result of the gradual deepening of the mixed layer and entrainment of hypolimnetic nutrients to the epilimnion. If latitude is low enough (e.g., less than 45 degrees) to allow for significant solar radiation in autumn and winter, then the combined effects of high light, relatively warm winter water temperatures (> 10°C) and winter mixing of nutrients from the hypolimnion allow for year round phytoplankton growth in such monomictic lakes.

In many New Zealand lakes, for instance, maximum phytoplankton biomass occurs in winter (Viner and White 1987,Vincent 1983) following a rapid autumnal growth (Fig. 2). Clearly nutrient reduction strategies for lakes that are based on minimising phytoplankton concentrations in summer, such as those derived from summer chlorophyll vs phosphorus concentration models (Sakamoto 1966, Dillon and Rigler 1974), may not be applicable to many south temperate monomictic lakes. In these conditions year-round nutrient additions to lakes including in winter underflows to the hypolimnion will need to be reduced to minimise winter accumulations of phytoplankton biomass. In contrast, the relationship of in-lake biomass and nutrient concentrations and loads in tropical lakes is often not clear or direct. Restoration of tropical lakes with high recycling rates and year-round high primary production needs to be directed towards biomass control, rather than direct control of nutrient inputs (von Sperling 1997).

As well as lake phytoplankton manipulations by nutrient load reductions, controls on excessive macrophyte growth by physical or chemical methods are also used in lake remediation programmes. Year round growth of macrophyte communities is a feature, not only in tropical lakes, but also of temperate lakes that are at latitudes that do not allow ice cover. For instance, lowland New Zealand lakes to latitude 47 degrees South have a year-round growth of littoral plant communities and introduced aquatic weeds thrive during all months. Thus local climatic conditions mean that excessive macrophyte weed control may involve cutting and harvesting of exotic weeds several times a year (Howard-Williams et al. 1996). However, even this method is subject to conditions on an individual lake scale. In one lake where strong currents were a feature of the littoral zone, restoration of native *Nitella* spp. communities in areas heavily invaded by exotic weeds was shown to be feasible after only one cut of the exotic species (*Lagarosiphon major*) (Howard-Williams et al. 1996). In an adjacent lake with weak water currents at least three cuts were needed each year.

Internal Loading — It is of interest that an almost universal finding has been that lakes respond slowly to initial nutrient reduction (e.g., Jeppesen et al. 1999, Madgwick 1999,

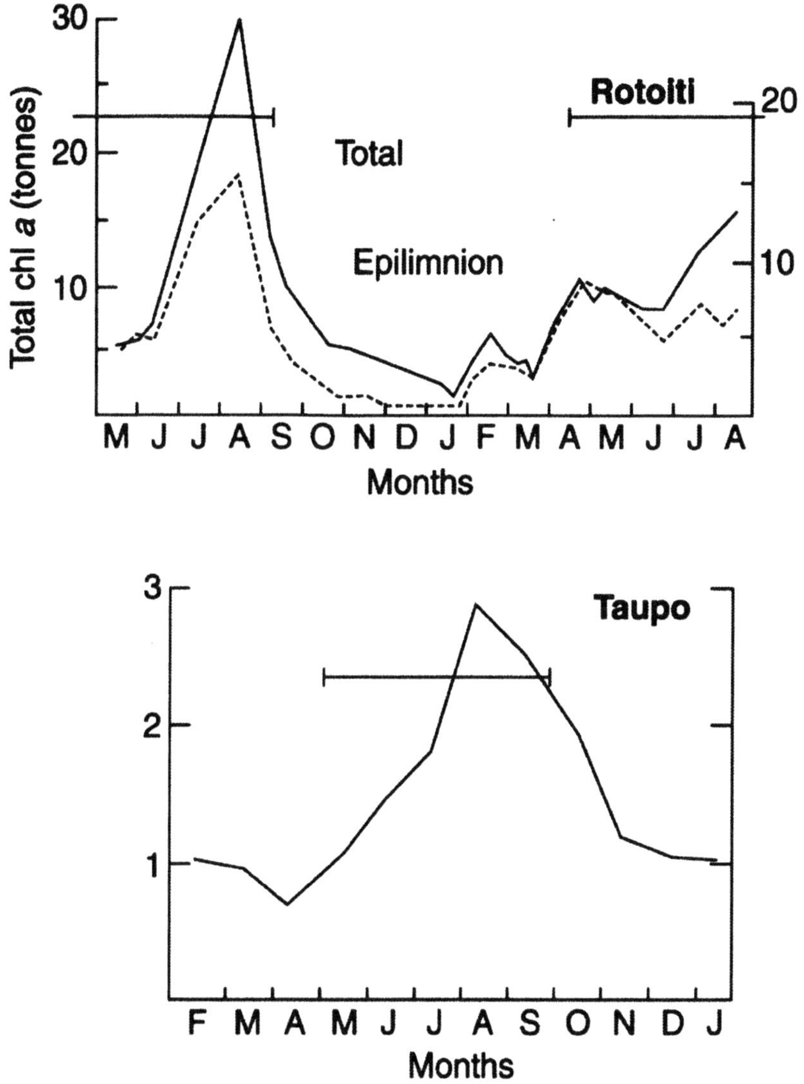

Fig. 2. Annual cycle of total phytoplankton biomass in two New Zealand north island lakes (Lakes Rotoiti and Taupo) showing development of phytoplankton biomass during the mixed period, reaching an annual maximum in winter. The mixed period is shown by a horizontal bar. The southern hemisphere winter is between June and September (Redrawn from Viner and White 1987).

Padisak and Reynolds, 1998, Parker and Maberley, 2000 and Vant and Gilliland 1991). The most common reason is that eutrophication has resulted in a rich sediment pool of nutrients, and internal loading has continued long after nutrient reduction in the catchment. The extent of internal loading is often climatically driven and therefore will vary on local scales. For Lake Rotorua in New Zealand, the internal load of dissolved inorganic N and P after four days of bottom deoxygenation was equivalent to the total annual load from all external sources. In this lake, conditions favouring bottom anoxia, and hence P-release, occur infrequently (sometimes once a year) and in an unpredictable manner dependent on local climatic conditions. Obviously, in such cases, restoration programmes will need to target the removal of sediment nutrient over long time-scales and the minimisation of the onset of anoxia.

Nurnberg (1984) provided a well-documented general model for the phosphorus retention coefficient for lakes (R). The relationship: $R = 15/(18+Q/A)$ where A is the lake surface area and Q is the hydraulic loading was found to hold for a wide variety of lakes with aerobic hypolimnia, including lakes in northern and southern hemispheres. The relationship breaks down as soon as anoxia occurs and internal loading reduces the net loss of P to the sediments. Vant and Hoare (1987) point out that the development of a generalised equation partitioning between internal and external phosphorus loads for anoxic lakes is not feasible due to the importance of local conditions that govern internal loading and that are specific to individual lakes.

Biogeography —One of the tenets of global models for remediation of eutrophication is the ability to alter the food web of the targeted system to some more desirable state, and to do this in a controlled predictable manner. The response of the system to reductions in point source and non-point source loadings of nutrients can vary considerably, and frequently may not result in the desired outcome. This has prompted considerable debate over the importance of top-down versus bottom-up controls on lake productivity (Carpenter et al. 1987, Jeppesen et al. 1990, Demelo et al. 1992, McQueen et al. 1989). As previously discussed, large differences in climate and geology between southern and northern temperate regions may further contribute to the lack of predictive capacity of these global models. Biomanipulation has been emphasized as an essential tool in restoring eutrophic systems where nutrient reductions alone may be insufficient, and top-down manipulation is necessary to drive the system (Van Donk and Gelati 1989, Jeppesen et al. 1990, Meijer et al. 1994, Klinge et al. 1995, Jeppesen et al. 1997a). However, it is also apparent from the literature that biomanipulation can often result in varying degrees of improvement in lake water quality, which are highly dependant upon the structure of the lake food-web (Irvine et al. 1989, Padisak and Reynolds 1998, McQueen 1998). This lack of definitive patterns to lake biomanipulations exemplifies the importance of the unique physio-chemical interactions with the food-web of a particular system, and the need for local perspectives in undertaking lake management. Most empirical evidence of these techniques has been derived for temperate lakes of Europe and North America (see reviews by Hosper 1998 and McQueen 1998), and may not be applicable in southern temperate regions where food-web structure greatly differs and lake faunas are species poor (Williams 1974, Frankenberg 1974, Forsyth and Lewis 1987, McDowall 1987). Below we look at fish communities, zooplankton, large aquatic plants, and phytoplankton separately.

Fish Communities — The first point we wish to raise is that native fish communities in many southern temperate lakes are such that the use of localised approaches to biomanipulation is essential. Biomanipulation typically involves either the enhancement of piscivorous fish, or the removal of herbivorus or planktivorous fishes with aims of enhancing zooplankton populations that graze on phytoplankton. Examples of this form of biomanipulation have been conducted predominantly in temperate northern hemisphere lakes with relatively complex food chains, containing several species of planktivorous fishes, planktivorous zooplankters, as well as piscivorous fishes (Fig. 3). Given the complexity of these systems, the degree of variability in response of these systems to the biomanipulation (Van Donk and Gelati 1989, McQueen 1998) is not surprising. However, in southern temperate lakes, food-webs are considerably simpler (Chapman et al. 1985, James et al. 2001) and lake faunas more impoverished than northern-hemisphere counterparts (McDowall 1990, Burns 1992). Although this may simplify potential responses to a given form of biomanipulation, it also limits the degree to which the system can be manipulated.

The example we cite is again for New Zealand, which has only 35 native freshwater fish species, of which only five regularly inhabit lakes (McDowall 2000) (Table 1). Most of these fish species are benthivorous as adults (although larval forms can be planktivorous), with the only two piscivorous species being eels (*Anguilla diffenbachii and A. australis*). There are no obligate planktivores, which is frequently the feeding guild targeted in lake biomaniupulation. It is also striking that there are no native herbivorous fishes, with the exception of the New Zealand grayling, which is now extinct. This relatively impoverished fauna is largely due to the young age of the landmass and its geographical isolation. All lake inhabiting species are not true lacustrine species, and have evolved from land-locked diadromous populations. By

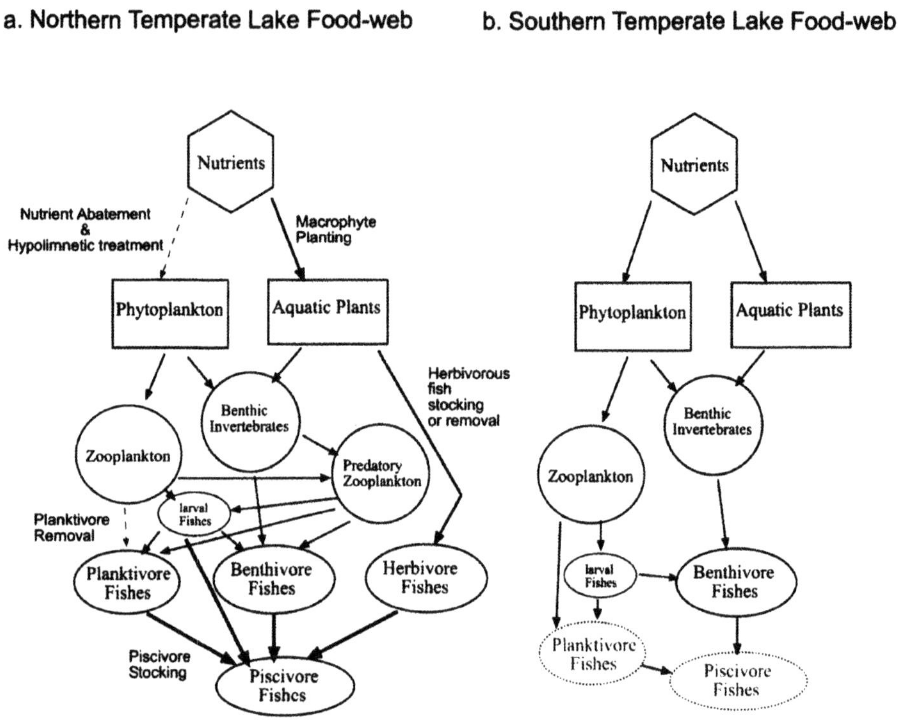

Fig. 3. Theoretical food-web energy transfer model for a) northern hemisphere and b) southern hemisphere temperate lakes. Arrows indicate energy flows between trophic levels with thick arrows indicating enhanced energy flows via biomanipulation (e.g., piscivore stocking) and dashed arrows indicating reduced energy transfer. Grey shaded trophic guilds (southern hemisphere only) represent non-native species only. Note that lake biomanipulations such as piscivore stocking or herbivorous fish stocking are only possible in northern hemisphere lakes if non-native fish species are not considered.

contrast, fish communities in temperate regions of North America (north of Mexico) have been estimated at approximately 212 species, composed of a multitude of feeding guilds (Lagler 1952). As such, the potential for biomanipulation of native fish communities in New Zealand lakes is somewhat limited, and would follow different strategies to models devised for North America and Europe. Likewise, fish faunas of southern Australia, Argentina and Chile tend to be low in diversity (Frankenberg 1974, Macchi et al. 1999) and methods for biomanipulation would be best drawn from local studies of food-webs than global restoration models.

Table 1. A list of all native New Zealand fish species that regularly inhabit lakes, along with their geographical distributions (in lakes) and diets. Note that no herbivorous species now exist, planktivorous forms generally only occur as larval stages, and the only piscivores are larger sized eels. Sourced from McDowall (2000).

Species	Lake distributions	Diet
Common bully (*Gobiomorphus cotidianus*)	widely distributed	benthivorous as adult, planktivorous as larva (<3 cm)
Common smelt (*Retropinna retropinna*)	mainly North Island	benthivorous and planktivorous
Koaro (*Galaxias brevipinnis*)	high elevation lakes	benthivorous, planktivorous as larva
Dune lakes galaxias (*Galaxias gracilis*)	localised	benthivorous and planktivorous
New Zealand grayling (*Prototroctes oxyrhynchus*)	extinct	herbivorous
Longfin eel (*Anguilla dieffenbachii*)	widely distributed	benthivorous as juvenile, piscivorous adult (>80cm)
Shortfin eel (*Anguilla australis*)	widely distributed	Benthivorous as juvnile, piscivorous as adult (>50 cm)

Introductions of exotic fish species may further complicate the ability to restore eutrophied systems in southern temperate lakes. The presence of exotic fishes in New Zealand lakes is pervasive, many to the demise of endemic populations (McDowall 1987, 1990, Crowl et al. 1992, Macchi et al. 1999). Salmonid species, mostly brown and rainbow trout, have been introduced throughout New Zealand, southern Australia and South America, and now serve as the top predator in many of these systems (Jackson 1981, McDowall 1990). Coarse fishes such as mosquitofish, rudd, tench, and European carp are also now widely distributed throughout north island lakes of New Zealand (McDowall 2000). Many of these systems to which coarse fish have been introduced (by and large illegally by misguided anglers) have undergone severe declines in water quality, and efforts are now being made to remove these populations and restore impacted systems. The options for biomanipulation of these invasive species are limited beyond completely extirpating the introduced species, as there are no native piscivores to keep populations under control. Introductions of grass carp (several locations throughout the North Island) and silver carp (Lake Omapere) have been permitted in instances where growths of aquatic weeds or excessive plankton blooms have resulted in severe deterioration of water quality. This type of biomanipulation is at risk, and is generally outside of the current political will in New Zealand. Many of the systems in which invasions have recently occurred are still in a state of flux, and predictive management by means of biomanipulation may prove difficult.

Zooplankton — Zooplankton communities in southern hemisphere temperate lakes are also indosyncratic by comparison to northern temperate systems (Chapman et al. 1985, Soto and Zuniga 1991), and may not be suited for global models of biomanipulation. This may be due largely to the idiosyncratic fish communities that have evolved in these systems. Similar to freshwater fishes, zooplankton communities of New Zealand tend to be very depauperate, and there are no large predatory planktivores like those that inhabit northern temperate lakes, such as *Leptadora* or *Chaoborus*. There is also a lack of native obligate planktivorous fishes in these systems, although some commonly occurring native species such as bullies and Koaro have planktivorous larvae (Table 1). Although it was originally thought that zooplankton populations were controlled mainly by bottom-up factors such as food supply and temperature in New Zealand lakes (Green 1976, Chapman et al. 1985), recent evidence suggests that top-down regulation may be important for shaping zooplankton community structure (Jeppesen et al. 1997b, 2000). Top-down regulation of zooplankton seems particularly apparent for systems where non-native fish species such as trout and perch have been introduced. Biomanipulation via removal of non-native fish biomass could be an effective tool for increasing herbivory by zooplankton, however, long-term maintenance of low planktivory by maintaining top-predator fish populations could not be done as it has in some Northern Hemisphere lakes because there are no native piscivores. As such New Zealand lakes present a unique set of challenges for biomanipulation strategies.

Aquatic Plants — Local models for aquatic plant management should also be favoured in southern temperate lakes, again due to the different native macrophyte communities. Often biomanipulation of upper trophic levels is combined with macrophyte planting to promote macrophyte dominance and improve water clarity. In New Zealand lakes, most native submersed aquatic plants are low-lying forms with slow growth rates (Howard Williams et al. 1987), and tall canopy forming plants that occur in European and North American lakes are absent. However, as with the fish fauna, introduced species of aquatic weeds remain a serious management concern throughout New Zealand, particularly tall-growing species such as *Lagarosiphon major* and *Egeria densa* which tend to outcompete native forms. Invasion by *E. densa* in some shallow eutrophic lakes in New Zealand has been shown to shift the system to alternating stable states macrophyte-phytoplankton dominance. In Lake Omapere, *Egeria* outcompetes native macrophytes to form dense monospecific stands, after which it undergoes a die-back stage during which phytoplankton productivity dominates in the system. Management of these species is further complicated by the fact there are no native herbivorous fish species in New Zealand, and only two species of invertebrates (freshwater Crayfish *Paranephrops planifrons* and the aquatic larva of the moth *Nymphula nitens*) are known to consume live macrophyte material. So again, the options for biomanipulation of submersed aquatic plants is more limited due to the exclusion of these components of the community, and emphasizes the importance for localized management decisions in favour of global models.

Phytoplankton — Most lakes that have been subjected to nutrient enrichment to the extent that they have become eutrophic, also show phytoplankton species shifts towards dominance by cyanobacteria. Lake Biwa is one of a worldwide series of examples. So in this respect there is a common global response. However, the reasons for cyanobacterial dominance vary markedly on local scales, and therefore, by inference the reduction in cyanobacterial blooms will require a local approach. On a lake-by-lake basis several factors have been implicated in cyanobacterial dominance in lakes (eg Vincent 1989; Varis, 1993, and many others). In addition to nutrient enrichment, these include: elevated water temperature, low N/P ratios, low light energy requirements, high pH and / low CO2 concentrations, ineffective digestion of many cyanobacteria by filter feeding zooplankton, suppression of zooplankton by toxic extra cellular metabolites from cyanobacteria, suppression of growth of competing algae also by extra cellular excretions, vertical migration, buoyancy mechanisms and the ability to form resting stages on anoxic sediment surfaces. These factors all tend to operate at different temporal scales, and in many cases are linked (Vincent 1989). In any restoration programme it needs to noted that some of these factors are controllable and some are not (Varis 1993), and in some cases, an effective change from cyanobacteria to another plankton dominant may be brought about by changing a related parameter such as dissolved oxygen levels. All of these factors have local connotations and generalised models predicting cyanobacterial blooms have not, for this reason, been successfully developed.

Varis (1993) points out that in real-life eutrophication management situations, there may be a number of factors and several groups of algae that need to be "managed" simultaneously. The question is "How do you do this?" A common approach in decision making may be to compare say two groups of algae operating under one factor and assess what might happen to one of these groups if the factor changed. With a situation of (say) four factors (n) and three groups of algae (p) the number of combinations for comparison is 12 (where $np(p\text{-}1)/2$) and this number grows very rapidly with every factor or algal group added (24 in the case of four groups of algae). Application of a global model to such situations is clearly difficult.

The Need for Local Emphasis in Lake Restoration

Viner and White (1987) pointed out that New Zealand lakes were sufficiently different from those of northern hemisphere European and North American lakes to require that water managers shift emphasis away from the simple alteration of mass nutrient budgets (especially phosphorus) for the control of eutrophication problems. They stressed the need for more precise ways of evaluating lake fertility. We recommend a further set of initiatives at a local scale; these are the use of adaptive management and of expert systems.

Adaptive management of lakes is the continuous changing of management techniques, and even the alteration of goals, to adjust to changes in the lake as a restoration process progresses. Adaptive management requires an understanding of the system and, in particular, a well structured monitoring programme so that changes can be detected and the direction of changes recognised. Varis (1993) recommends the use of analytical computational tools for lake management in the form of expert systems. We believe these should be applied on a case-by-case basis with special emphasis on the inclusion of regional and local knowledge on controls on lake fertility. Restoration may be very haphazard without this consideration.

It could be argued that the Netherlands is an example where a "global" approach to lake restoration has been relatively successful. However, the relative success of the restoration programmes in the Netherlands (Portielje and Van der Molen 1999) is because in a global context the landscape, lake types and reasons for lake fertility in the Netherlands are (almost) homogeneous, so a single approach (phosphorus removal in stages (see above)) has been relatively successful. Even here there are examples where unique sets of local conditions have combined with the result that the restoration programmes have not always worked. A true "global" approach would entail translating successful national programmes (such as those in the Netherlands) to, say, mountainous countries with maritime climates and volcanic soils. This is obviously not a reasonable approach.

In truth, the real solution lies between local and global arguments. The concept of resilience is fundamental to understanding restoration and the long-term sustainability of "restored"`lakes. Resilience is an ecological paradym of global applicability and the importance of resilience in lake restoration has recently been eloquently addressed by Carpenter and Cottingham 1997. Within this global "model", each lake, and especially each human community with interest in that lake, have specific controlling forces that determine its restoration potential and its pathway to a restored state. In particular it is the local human community that has to decide on what is a "restored" system (this probably will not be the original pre-degraded lake). This decision is intensely "local" and depends on the cultural background of the community, its ability to afford restoration and its plans for the use of the lake into the future. Carpenter and Cottingham (1997) point out that concepts that integrate people and lakes must consider the processes that control normal and degraded lakes. Restoration is costly, and their needs to be a political will with appropriate recognition of the process and its funding needs. The community should at least acknowledge the processes that govern lakes and how the restoration process will alter these.

Nowhere has this been better described than for a small lake in Wales where conflicting community and regulatory authority interests and a lack of understanding of the ecological processes involved has resulted in the almost complete failure of a restoration programme (Moss et al. 2002). Scientific advice was followed initially, with early successes, but as the project continued, local competing financial and political interests became involved, recreation interests that did not support the project interfered, and there were problems in communication.

The prime lesson from this restoration project according to the analysis of Moss and his colleagues was that the target for restoration should not only be carefully determined but that it must be supported by all concerned in the local community. This, combined with regional scale factors described in this chapter involving local geology, climate, landscape and biodiversity, mean that lake restoration projects can only be conducted on local scales.

Acknowledgements

Professor Michio Kumagai is thanked for the invitation to attend the international workshop on Global versus Local Perspectives of Water Management in Otsu Japan 2001, and for funding travel expenses. Workshop participants, in particular Dr Erik Jeppesen are thanked for stimulating discussions on this topic. The New Zealand Foundation for Research Science and Technology is acknowledged for funding under Contract CO1X0020, and the National Institute for Water and Atmospheric Research is thanked for providing additional time for this project. Helpful comments on this manuscript were provided by Dr. Bob McDowall and Dr. Ian Hawes and an anonymous reviewer.

References

BURNS, C. W. 1991. New Zealand Lakes Research, 1967-91 N. Z. J. Mar. and Fresh. Res. **25:** 359-379.

________. 1992. Population dynamics of crustacean zooplankton in a mesotrophic lake, with emphasis on *Boeckella hamata* Brehm (Copepodoa:Calanoida). Internat. Rev. der ges. Hydrob. **77:** 553-577.

CARPENTER, S. R., J. F. KITCHELL, J. R. HODGSON, AND K. L. COTTINGHAM. 1987. Regulation of lake primary productivity by food-web structure. Ecology **68:** 1863-1876.

________, AND K. L. COTTINGHAM. 1997 Resilience and restoration in lakes. Cons. Ecol. Online **1**(1): 1-18

________, D. BOLGRIEN, R. C. LATHROP, C. A. STOW, T. REED, AND M. A. WILSON 1998. Ecological and economic analyses of lake eutrophication by non-point pollution. Aust. J. Ecol. **23:** 68-79.

________, AND R. C. LATHROP. 1999. Lake Restoration: capabilities and needs. *In* D. M. Harper, B. Brierley, A. J. D. Ferguson and G. Phillips [eds.], The Ecological Bases for Lake and Reservoir Management. Hydrobiologia **395/396:** 19-28.

CHAPMAN, M. A., J. D. GREEN, AND V. H. JOLLY. 1985. Relationships between zooplankton abundance and trophic state of seven New Zealand lakes. Hydrobiologia **123:** 119-136.

________, AND J. D. Green. 1987. Zooplankton Ecology. p. 225-263. *In* A. B. Viner [ed.], Inland Waters of New Zealand. DSIR Science Information Publishing Center, Wellington, New Zealand.

COLLIER, K., A. B. COOPER, R. J. DAVIES-COLLEY, J. C. RUTHERFORD, C. M. SMITH, AND R. B. WILLIAMSON. 1995. Managing Riparian Zones: A contribution to protecting New

Zealand streams. Volume 2: Guidelines. Department of Conservation, Wellington, NZ, 142 p.

CROWL , T. A., C. R. TOWNSEND, AND A. R. MCINTOSH. 1992. The impact of introduced brown and rainbow trout on native fish: the case of Australasia. Rev. Fish Biol. Fish **2:** 217-24.

DEMELO, R., R. FRANCE, AND D. J. MCQUEEN. 1992. Biomanipulation: Hit or myth? Limnol. Oceanogr. **37:** 192-207.

DEVITO, K. J., I. F. CREED, R. L. ROTHWELL, AND E. E. PREPAS. 2000. Landscape controls on phosphorus loading to boreal lakes: impliucations for potential impacts of forest harvesting. Can. J. Fish. Aquat. Sci. **57:** 1977-1984.

DILLON, P. J. AND F. H. RIGLER. 1974. A test of a simple nutrient budget model predicting the phosphorus concentration in lake water. J. Fish. Res. Board Can. **31:** 1771-1778.

DOWNES, M. T., C. HOWARD-WILLIAMS AND L. SCHIPPPER. 1997. Long and short roads to riparian zone restoration: nitrate removal efficiency p. 244-253. *In* N. E. Haycock, T. P. Burt, K. W. T. Goulding and B. Pinay [eds.], Buffer Zones: Their processes and Potential in Water Protection. Quest Environmental Publishers, Harpendon U.K.

FRANKENBERG, R. 1974. Native freshwater fish. p. 113-170. *In* W. D. Williams [ed.], Biogeography and Ecology in Tasmania. Dr. W. Junk Publishers, The Hague, Netherlands.

FORSYTH, D. J., AND M. LEWIS. 1987. Zoogeography: The invertebrates. p. 265-290. *In* A. B. Viner [ed.], Inland Waters of New Zealand. DSIR Science Information Publishing Center, Wellington.

GIBBS, M. M., AND F. E. MATHESON. 2001. Lake-edge wetlands and their importance to the Rotorua lakes, p. 99-106. *In* Proceedings and report on a symposium on Research Needs of the Rotorua Lakes, Lakes Water Quality Society, Rotorua, New Zealand.

GOLDMAN, C. R., AND A. J. HORNE. 1983. Limnology. McGraw-Hill, New York

GOLTERMAN, H. L. 1975. Physiological Limnology: An approach to the physiology of lake ecosystems. Elsevier Scientific Publishing Co. Amsterdam, 489p.

GREEN, J. 1976. Population dynamics and production of the calanoid copepod *Calamoecia lucasi* in a northern New Zealand lake. Archiv fur Hydrobiol., Supplement **50:** 313-400.

HARPER, D. M., B. BRIERLEY, A. J. D. FERGUSON, AND G. PHILLIPS, G. [eds.], 1999. The Ecological Bases for Lake and Reservoir Management. Hydrobiologia **395/396:** 469p.

HAWES, I., AND R. SMITH. 1993. Effect of localised nutrient enrichment on the shallow epilithic periphyton of oligotrophic Lake Taupo, New Zealand. N. Z. J. Mar. and Fresh. Res. **27:** 365-372.

HOWARD-WILLIAMS, C. 1986. Studies on the ability of a *Potamogeton pectianatus* community to remove dissolved nitrogen and phosphorus compounds from lake water. J. Applied Ecol. **18:** 619-637.

________, J. S. CLAYTON, B. T. COFFEY, AND I. M. JOHNNSTONE. 1987. Macrophyte invasions, p. 307-331. *In* A. B. Viner [ed.], Inland Waters of New Zealand. DSIR Science Information Publishing Center, Wellington.

________, A-M. SCHWARZ, AND V. REID. 1996. Patterns of aquatic weed regrowth following mechanical harvesting in New Zealand hydro-lakes. Hydrobiologia **340:** 229-234.

HOSPER, S. H. 1998. Stable states, buffers and switches : an ecosystem approach to the restoration and management of shallow lakes in the Netherlands. Wat. Sci. Tech. **3:** 151-164.

IRVINE, K., B. MOSS, AND H. BALLS. 1989. The loss of submerged plants with eutrophication II. Relationships between fish and zooplankton in a set of experimental ponds, and conclusions. Fresh. Biol. **22:** 89-107.

JACKSON, P. D. 1981. Trout introduced into south-eastern Australia: their interaction with native fishes. Vict. Natural. **98:** 18-24.

JAMES, M. R., I. HAWES, M. WEATHERHEAD, C. STANGER, AND M. GIBBS. 2001. Carbon flow in the littoral food web of an oligotrophic lake. Hydrobiologia **441:** 93-106.

JEPPESEN, E., M. SONDERGAARD, E. MORTENSEN, P. KRISTENSEN, B. RIEMANN, H. J. JENSEN, J. P. MOLLER, O. SORTKJAER, J. P. JENSEN, K. CHRISTOFFERSEN, S. BOSSELMANN, AND S. DALL. 1990. Fish manipulation as a lake restoration tool in shallow, eutrophic temperate lakes. 1. Cross-analysis of three Danish case-studies. Hydrobiologia **200/201:** 205-218.

________, J. P. JENSEN, M. SONDERGAARD, T. LAURIDSEN, L. JUNGE, AND L. JENSEN. 1997a. Top-down control in fresh-water lakes: the role of nutrient state, submerged macrophytes and water depth. Hydrobiologia **342/343:** 151-164.

________, T. LAURIDSEN, S. F. MITCHELL, AND C. W. BURNS. 1997b. Do planktivorous fish structure the zooplankton communities in New Zealand lakes. N. Z. J. of Mar. and Fresh. Res. **31:** 163-173.

________, T. L. LAURIDSEN, S. F. MITCHELL, K. CHRISTOFFERSEN, AND C. W. BURNS 2000. Trophic structure and the pelagial of 25 shallow New Zealand lakes: changes along nutrient and fish gradients. J. Plank. Res. **22:** 951-968.

________, M. SONDERGAARD, B. KRONGVANG, J. P. JENSEN, L. M. SVENDSEN, AND T. L. LAURIDSEN. 1999. Lake and catchment management in Denmark. Hydrobiologia **395/396:** 419-432.

JIN, X. 1994. An analysis of lake eutrophication in China. Mitt. Internat. Verein. Limnol. **24:** 207-211.

JOHN, P. H., M. M. GIBBS, AND M. T. DOWNES. 1978. Groundwater quality along the eastern shores of Lake Taupo 1975-1976. N. Z. J. of Mar. and Fresh. Res. **12:** 59-66

KLINGE, M., M. P. GRIMM, AND S. H. HOSPER. 1995. Eutrophication and ecological rehabilitation of Dutch lakes: presentation of a new conceptual framework. Wat. Sci. Tech. **8:** 207-218.

KRANTZ, T. K., K. E. WEBSTER, C. J. BOWSER, J. J. MAGNUSON, AND B. J. BENSON. 1997. The influence of landscape position on lakes in northern Wisconsin. Fresh. Biol. **37:** 209-217.

LAGLER, K. F. 1952. Freshwater Fishery Biology. Wm. C. Brown Company Publishers, Dobuque, Iowa, USA. p. 19-61.

MCDOWALL, R. M. 1987. Impacts of exotic fishes on native fauna. P. 291-306. *In* A. B. Viner [ed.], Inland Waters of New Zealand. DSIR Science Information Publishing Center, Wellington.

________. 1990. Filling in the gaps – the introduction of exotic fishes into New Zealand. p. 68-92. *In* D. A. Pollard [ed.], Introduced and Translocated Fishes and their Ecological Effects. Proceedings of the 8th Australian Society for Fish Biology Workshop, Magnetic Island, 24-25 August 1989.

________. 2000. The Reed Field Guide to New Zealand Freshwater Fishes. Reed Publishing, Auckland. 224p

MACCHI, P. J., V. E. CUSSAC, M. F. ALONSO, AND M. A. DENEGRI. 1999. Predation relationships between introduced salmonids and the native fish fauna in lakes and reservoirs in northern Patagonia. Ecol. Fresh. Fish **8:** 227-236.

MCQUEEN, D. J., M. R. S. JOHANNES, J. R. POST, T. J. STEWART, AND D. R. S. LEAN. 1989. Bottom-up and top-down impacts on freshwater pelagic community structure. Ecol. Monog. **59:** 289-309.

________. 1998. Freshwater food web biomanipulation: A powerful tool for water quality improvement, but maintenance is required. Lakes and Reservoirs: Res. and Manage. **3:** 83-94.

MADGWICK, F. J. 1999. Restoring nutrient-enriched shallow lakes: integration of theory and practice in the Norfolk Broads, U.K. Hydrobiologia **408/409:** 1-12

MEIJER, M-L. E. JEPPESEN, E. VAN DONK, B. MOSS, M. SCHEFFER, E. LAMMENS, E. VAN NES, J. A. BERKUM, G. J. DE JONG, B. A. FAAFEENG, AND J. P. JENSEN. 1994. Long-term responses to fish stock reduction in small shallow lakes: Interpretation of five year results of four bio-manipulation cases in the Netherlands and Denmark. Hydrobiologia **275/276:** 457-466.

MOSS, B., L. CARVALHO, AND J. PLEWES. 2002. A lake at Llandrindod Wells – a restoration comedy? Aquat. Cons. Mar. and Fresh. Syst. **12:** 229-245.

NURNBERG, G. K. 1984. The prediction of internal phosphorus load in lakes with anoxic hypolimnia. Limnol. Oceanogr. **29:** 111-124.

OECD. 1982. Eutrophication of waters – monitoring, assessment and control. OECD, Paris. 154 p.

OHLE, W. 1982. Nährstoffzufuhren des Grebiner Sees durch atmosphärische Niederschläge und Oberflächenabschwemmung des Einzuigsgebietes. Arch Hydrobiol. **95:** 331-363.

PADISAK, J., AND C. S. REYNOLDS. 1998. Selection of phytoplankton associations in Lake Balaton, Hungary, in response to eutrophication and restoration measures, with sprcial reference to the cyanoprokaryotes. Hydrobiologia **384:** 41-53

PARKER, J. F., AND S. C. MABERLEY. 2000. Biological response to lake remediation by phosphate stripping:control of *Cladophora*. Fresh. Biol. **44:** 303-309.

PHILLIPS, G., A. BRAMWELL, J. PITT, J. STANSFIELD, AND M. PERROW. 1999. Practical application of 25 years' research into the management of shallow lakes. Hydrobiologia **396/396:** 61-76.

PORTIELJE, R., AND D. T. VAN DER MOLEN. 1999. Relationships between eutrophication variables: from nutrient loading to transparency. Hydrobiologia **408/409:** 375-387.

SAKAMOTO, M. 1966. Primary production by phytoplankton community in some Japanese lakes and its dependence on depth. Arch. Hydrobiol. **62:** 1-28.

SONDERGAARD, M., E. JEPPESEN, AND J. P. JENSEN. 2000. Hypolimnetic treatment to reduce internal phosphorus loading to a stratified lake. Lake and Reser. Manage. **16:** 195-204.

SOTO, D., AND L. ZUNIGA. 1991. Zooplankton assemblages of Chilean temperate lakes: A comparison with North American counterparts. Revista Chilena de Historia Natural **64:** 569-581.

VANT, B., AND B. HUSER. 2000. Effects of intensifying land-use on the water quality of Lake Taupo. Proc. N. Z. Soc. of Anim. Prod. **60:** 261-264.

________. 1987a. Eutrophication: An overview. *In* W. N. Vant [ed.], Lake Managers Handbook. National Water and Soil Conservation Authority, Wellington. Wat. and Soil Misc. Pub. **103:** 151-157.

________. 1987 b. Lake Managers Handbook. National Water and Soil Conservation Authority, Wellington. Wat. and Soil Misc. Pub. **103:** 230p.

________, AND B. W. GILLILAND 1991. Changes in water quality in Lake Horowhenua following sewage diversion. N. Z. J. Mar. Fresh. Res. **25:** 57-61.

________, AND R. A. HOARE. 1987. Determining the input rates of plant nutrients *In* W. N. Vant [ed.], Lake Managers Handbook. National Water and Soil Conservation Authority, Wellington. Wat. and Soil Misc. Publ. **103:** 151-157.

VAN DONK, E., AND R. D. GELATI. 1989. Biomanipulation in the Netherlands. Proceedings of a symposium. Hydrobiol. Bull. **23**: 1-99.

VARIS, O. 1993. Cyanobacteria dynamics in a restored Finnish Lake: Long term simulation study. Hydrobiologia **268:** 129-145.

VINCENT, W. F. 1983. Phytoplankton production and winter mixing: contrasting effects in two oligotrophic lakes. J. Ecol. **71:** 1-20.

________. [ed.], 1989. Cyanobacterial growth and dominance in two eutrophic lakes. Ergebnisse der Limnologie (Special Issue). **32:** 1-254.

VINER, A. B., AND E. W. WHITE. 1987. Lake Plankton processes: Phytoplankton growth. p. 191-223. *In* A. B. Viner [ed.], Inland waters of New Zealand. DSIR Science Information Publishing Center, Wellington.

VOLLENWEIDER, R. A. 1979. Das Nährstoffbelastungskonzept als Grundlage für den externen Eingriff in den Eutrophierungsprozess stehender Gewässer und Talsperren. Z. Wass Abwass. Forsch. **2/79:** 45-56.

VON SPERLING, E. 1997. The process of biomass formation as the key point in the restoration of tropical eutrophic lakes. Hydrobiologia **342/343:** 351-354.

WHITE, E. W. 1982. Eutrophication in New Zealand lakes, p. 74-78. *In* Water in New Zealand's Future; Proceedings of the 4th National Water Conference. Institution of Professional Engineers of New Zealand, Auckland.

________. 1983. Lake eutrophication in New Zealand – a comparison with other countries of the Organisation for Economic Co-operation and Development. N. Z. J. Mar. and Fresh. Res. **17:** 437-444.

________, G. PAYNE, S. PICKMERE, AND P. WOODS. 1986. Nutrient demand and availability related to growth among natural assemblages of phytoplankton. N. Z. J. Mar. and Fresh. Res. **20:** 199-208.

WILLIAMS, W. D. 1974. Freshwater Crustacea. p.113-170. *In* W. D. Williams [ed.], Biogeography and Ecology in Tasmania. Dr. W. Junk Publishers, The Hague, Netherlands.

Chapter 6

Requirements for Lake Management

6-1. Strategies for Lake Management in an Increasingly Global Environment

Charles R. Goldman

Director, Tahoe Research Group, Department of Environmental Science and Policy, University of California, Davis, CA 95616

Abstract

Fresh water, the most essential resource, is in increasingly short supply globally and is likely to be the cause of future conflicts. As an essential but seriously undervalued "commodity", it should take its place among the other well-known commodities and be better managed by world governments. Aquatic ecosystems worldwide are under increasing anthropogenic stress. This situation necessitates a more rapid conversion of scientific studies into effective management decisions. Atmospheric-borne pollution of the world's ecosystems demonstrates the importance of achieving a global perspective as we are forced to face the growing challenge of declining environmental quality. The conservation of lakes and streams as well as the protection of drinking water sources is now of urgent concern. The continuing loss of clarity in Lake Tahoe provides a vivid example of the problems being faced globally. Lake Tahoe is losing its remarkable Secchi transparency at an annual rate of 0.3 meters as algal growth rates increase concomitantly. A multidisciplinary approach has been essential to developing effective management strategies at Tahoe and elsewhere for solving increasingly complex environmental problems. Long-term data collection, including paleolimnological studies of sedimentation and pollutants, has been key to better understanding and managing the lake, its surrounding watershed, and basin air quality. Previously, many policy decisions by regulatory agencies were based on scanty short-term data that were methodologically lacking or subject to superficial interpretation. The latter case was exemplified by a brief drought-related improvement in transparency at Tahoe. Educating the public and their political leadership is increasingly a very important task for the scientific community. Modern ecologists and limnologists have a responsibility to help meet the growing global challenge for restoration and preservation of threatened water supplies. Strong environmental science based on long-term studies must be at the forefront in developing improved adaptive management practices for both aquatic and terrestrial ecosystems worldwide.

Introduction

Water as a Global Commodity — A somewhat different perspective may be useful as we complete our discussions of the many important papers presented at this conference and look toward future needs with a global viewpoint. In general we are painfully aware that world water supplies are over-exploited, mismanaged and, above all, undervalued. There is an increasing need to improve management of water resources to retain the natural beneficial functions of ecosystems everywhere. Who would have believed a few years ago that we would now be paying more for "designer" water than for the same volume of gasoline? This paper first presents a proposal to re-think our overall valuation of world water supplies and treat them as a *commodity* rather than as a simple exploitable resource. Since climatic change now has all the appearance of reality and scientific acceptance, a global approach to anticipating its potential effects on the physical and chemical properties as well as the biology of surface waters will also be considered. Finally, atmospheric pollution and global transport will be discussed not only as the negative reality of fertilization and pollution, but also as a means of bringing the world community together to face this growing threat. The importance of involving political figures and the general public using easy-to-understand graphics is also noted. Since most of the author's day-to-day experience in research, teaching and publication has dealt with Lake Tahoe and Castle Lake, California, the lessons learned there will often be referred to in the context of our still-unresolved world water problems.

The water for irrigated agriculture is already to some extent considered a commodity in the same sense as oil, coffee, pork bellies, and orange juice. Our thinking probably should be readjusted to managing our declining world water supplies, particularly those suitable for drinking, as the valuable commodity they represent. After all, water is often sold in the U.S. for agricultural use at a certain price per acre-foot. Irrigated agriculture, of course, requires that the supply and demand market determine the cost of water purchases. Because of the importance of agriculture to so many economies, however, the price is often highly subsidized by governments. This tends to obscure its real value for other uses. By considering fresh water as a world commodity, available to all through fair international agreements, we should be better able to husband this essential resource for life as we know it. In South Africa, prioritization of water supplies has led to serious mismanagement and the worst cholera epidemic in the country's history. If it has to be transported too far or if not enough is used in irrigation, the soil soon loses the capacity to grow crops due to increasing soil salinity. It is obvious that the economics of water management really drive the use and misuse of this renewable but not totally recoverable resource. We are all aware that civilizations of the past have perished due entirely to the misuse of this essential commodity. An ecological perspective is now finding its way into more management decisions, as noted by the late W.T. Edmondson in his book The Uses of Ecology (Edmondson 1991).

Atmospheric Pollution as a Globalizing Influence — From the perspective of utilization of water, it is interesting to recognize that atmospheric pollution is actually a globalizing influence on our environmental considerations. Acid rain originating from the Ruhr Valley industries, the British Isles, and the industrialized Midwest of the United States was recognized as the cause of the acidification of many lakes during the 20th Century. This problem was particularly well studied and recognized in the poorly buffered lakes of Scandinavia. High-sulfur coal burned in China will continue to have an impact throughout much of Asia and the world. In 2002, drought-caused dust borne on the global winds obscured the sun in parts of China and Korea and was transported as far as North America. We have long recognized the fact that industrial nations, particularly the United States, Canada, Japan, Germany and many other European countries, are now burning and for many years have burned most of the fossil fuel that is still being consumed at alarming rates. These countries are contributing a high percentage of the greenhouse gases and atmospheric pollutants that contribute to global warming and climatic change. We are beginning to appreciate the fact that our aquatic resources, together with terrestrial areas, are impacted by changes in atmospheric conditions, whether natural or anthropogenic. Deforestation in tropical regions alters albedo, reduces or eliminates a high percentage of transpiration, and can drastically reduce annual rainfall. Nitrogen pollution from atmospheric fallout has already altered the nutrient dynamics of Lake Tahoe (Goldman 2000a), as well as that of nonagricultural lands. Mercury has become an ubiquitous pollutant of waters in the U.S. and other industrial nations. The well-known Minamata Bay tragedy in Japan brought world attention to this danger which, through atmospheric transport, adds to our global understanding of the problem.

Planning for Climatic Change — With the growing threats from the undesirable effects of climatic change, the depletion of atmospheric ozone and the increase in Ultraviolet B radiation, particularly in the Southern Hemisphere, entire nations must reconsider their options for maintaining or improving the quantity and quality of their water supplies. Modeling has the potential of becoming a valuable predictive tool, but will often require decades to be validated. In the meantime, the scant supply of the world's usable surface and underground water commodity will continue to be at the mercy of these as-yet uncertain climatic impacts. Eutrophication has been with us long enough to be known as a real and continuing threat to the quality of surface waters. The prospect of major shifts in rainfall, bringing more to some areas and less to others, has already been brought to the attention of world leaders during the 2002 World Conference on Sustainability held in South Africa. Much importance was given at the conference to the water-short and water quality-poor regions of the world that have experienced high infant mortality and great human misery and starvation due to water-deficient failing crops. What was dealt with in only a small subsection of this huge conference was the more subtle effects of climatic change on the functioning of surface waters and the enormous challenges of meeting the world's growing needs. Lost in the quest for sustainability of development, the ongoing and escalating problems of unchecked population growth were left largely to the imagination of the delegates.

California, like many regions of the world, has had water shortages and problems for the last century. Reservoirs dot the western slopes of the Sierra Nevada range and the great Shasta dam in northern California collects and stores water from the Sacramento River. To sustain California, the sixth largest economy in the world, aqueducts have been required to carry water south from the relatively water-rich north part of the State, and the Colorado River had to be tapped to make up most of the remaining shortfall for the growing urban population and the important irrigated agriculture of southern California. Various schemes have been proposed to deliver more water from as far north as Oregon and Alaska. Most recently, the re-flooding of currently farmed delta islands is being considered to store the more abundant flows of winter and spring. These islands had originally been diked off from the Sacramento River to create additional fertile farmland. Studies are under way to evaluate the potential consequences of this new storage scheme. The rate of carbon loading from decomposition of peat deposits on the muddy bottom may be a major obstacle to using the islands for water storage. Since the water would be destined for municipal use, the cost of treatment is of major concern. Throughout the world, the costs of treating water increase precipitously if the quality of the source water declines. As example, with the Olympic Games destined for Beijing, China, finding treatable source water for the multiple venues will be a serious challenge. Eutrophication remains a major cause of source water degradation; therefore, on the global front, controlling eutrophication is an increasingly urgent and logical approach that will continue to depend on application of the best available science.

The specter of climatic change in association with global warming has a variety of potential impacts on our surface waters beyond the concern for rising sea level and redistribution of rainfall. As the surface waters of a lake are warmed, their resistance to mixing will increase. Temperate lakes will become more like tropical lakes, and far-northern lakes will become more like temperate lakes. This increased resistance to mixing or overturn will shorten or even eliminate the period of re-oxygenation of the deep waters and will cause reduced conditions over the sediment layer in many lakes. The reduced sediments will then be able to release their iron and associated phosphorus nutrients to the lake water, causing the well-known phenomenon of internal nutrient loading. As a result, there will be an increase in trophic status of many lakes as they become more eutrophic due to the warmer temperatures and increased nutrient supply. Although increasing algal growth in the near-surface waters, loss of deeper waters to anoxic conditions will reduce or eliminate benthic productivity and fish habitat.

The Lake Tahoe Example

Long-term data collection is extremely important. At Lake Tahoe and Castle Lake, California, it has made it possible to compare the behavior of a small lake with short retention time with the larger Tahoe that has a 700-year retention time (Goldman and de Amezaga 1984). Heavy rain and snowfall tends to reduce productivity of the short retention time Castle Lake, while the same conditions tend to provide deep mixing of Tahoe with resultant higher productivity. Further, the 30-year database for Castle Lake allowed us to predict lake warming on the basis of the various existing models of carbon dioxide-driven global warming (Byron and Goldman 1990). These concerns are applicable to lakes everywhere. Advances in science and technology may be capable of reducing some of the likely impacts, but a universal attack on eutrophication and causes

Fig. 1. Lake Tahoe as viewed from 5,300 meters above sea level looking towards the southeast. The lake has a surface area of 500 square kilometers and an average depth of 313 meters. It is the eleventh deepest lake in the world and contains 156 cubic kilometers of extremely high quality water. The lake is in the earliest stages of eutrophication and has been under extensive limnological study since 1959.

of global warming should be the first line of defense.

Another case of emerging water problems that are global in nature is also illustrated by Lake Tahoe, located at the crest of the Sierra Nevada between the states of California and Nevada (Figure 1). This lake is a well-known example of dealing with water problems as it has undergone extraordinary change since monitoring began in 1959 (Goldman 2000a). The lake is renowned as a tourist attraction for its cobalt-blue waters, is the main source of water for Truckee, California, the desert community of Reno, Nevada, Pyramid Lake, and irrigated agriculture to the east. Dealing with its problems has involved two states, five counties and various federal and state agencies. To coordinate efforts to save the lake, it was first necessary to create a regional planning agency that had a federal mandate to control development and begin restoration of this valuable natural resource. This took place in the early 1960's when the League to Save

Lake Tahoe, with the author's presence and scientific data documenting the early stages of eutrophication, was successful in convincing the governors of both states that a bi-state agency, the Tahoe Regional Planning Agency, known now as the TRPA, needed to be formed immediately to regulate further development and better manage the Tahoe basin with federally-mandated enforcement power. Once TRPA was in place, it was possible to establish new regulations and begin to repair the damage already done to the watershed. A sewage collection system was installed around the entire lake in order to treat and export wastewater from the basin. Shortly thereafter, solid wastes were also exported from the basin. In recent years, the need for management-driven research has become increasingly apparent (Goldman 2000b).

Lake Tahoe is also a particularly good example of the impact of heavy atmospheric nitrogen deposition that over a period of thirty years has changed this classically nitrogen-limited system to an extremely phosphorus-sensitive one. Largely due to atmospheric deposition, the ratio of nitrogen to phosphorus has changed from 1:1 to 40:1. In considering the global nature of world water supplies, they are all being altered to various extents by atmospheric pollution and global warming. For example, in Russia, through the diversion of one commodity, water, which was originally destined for the Aral Sea to supply another commodity, cotton, the diversion has already destroyed the extremely valuable fishery resource and put thousands of fishermen and cannery workers out of work. This commodity tradeoff, which was unanticipated, has had additional dire consequences. This great lake has been deprived of its inflow and has receded many kilometers from its original shoreline. The exposed littoral zone of the former sea has dried into a dusty desert that is constantly eroded by the winds (Goldman 1994). This dust has caused an epidemic of respiratory distress in the local populations. Traces of the wind-borne dust can now be found deposited as far north as the Arctic Circle. Globalization of environmental problems is thus a reality, through air transport of factory soot, automobile and aircraft exhaust, fertilizers, pesticides, and wind-borne contaminants from the deserts and dried lakebeds of the world.

Clean Water Supplies and World Peace

Future sources of clean water supplies remain a critical issue to achieving world peace. It is realistic to consider that future conflicts are as likely to arise over water supplies as they are over oil. The Jordan River out of Syria is the source of water for the Sea of Galilee, which supplies water to Israelis and Palestinians that have been locked in conflicts for decades. Current droughts in Africa and other regions of the world are already causing great human misery and increased infant mortality. Countries throughout the world are increasingly looking for new sources of water. This usually involves constructing aqueducts to move water great distances or depleting existing groundwater supplies. In past years, consideration was actually given to towing large icebergs from Antarctic waters to warmer latitudes. An international conference held at Ames, Iowa, in the U.S. was organized and supported by Prince Faisal of Saudi Arabia to consider the prospects. Since icebergs do not survive in the warmer oceans to the north, this author proposed a non-towing solution (Goldman 1978). Why not use obsolete single-hulled supertankers to move crushed ice from the poles? The ancient Antarctic ice has the value of low salinity, which makes it useful for irrigation, and low contamination. Further, its low temperature could be utilized for thermal electric power generation from the warmer surface waters of the tropics or as cooling water for urban air conditioning. It is, in the author's opinion, particularly unfortunate that this idea was abandoned along with the impractical, if not impossible, plans for towing icebergs. Another practical approach is to collect fresh water from large rivers just before it enters the sea. This can only be practical if the water has not been excessively contaminated by upstream pollution.

The Need for Global Water Monitoring

If a global network of monitoring stations is to be developed during and following the World Water Forum 3, it will be necessary to utilize low-cost, easy-to-maintain instrumentation and technology in order to achieve significant worldwide coverage. Relatively inexpensive water monitoring kits could be an important feature of this program. A favorite limnological instrument is the Secchi disc, which has no moving parts or electronics and should be included in any monitoring kit. This simple instrument for measuring lake and ocean transparency was first used aboard the Pope's Vatican Naval ship S.S. *l'Immacolata Concezione* along the Adriatic Sea by a Jesuit priest, Father Pietro Angelo Secchi, in 1865. He recognized the importance of water transparency in identifying submerged coastal hazards such as rocks or chain used to protect harbor mouths from uninvited ships. The data derived from Secchi measurements compares very well with transparency measures by the more sophisticated but often less reliable data collected with expensive submersible spectrophotometers. These measures of transparency have great value in that they mirror changes in trophic status, algal growth, and sediment and dust inputs that cloud the waters of the world's lakes. For example, the Secchi data collected over 35 times per year at Lake Tahoe has served very well over the years in communicating the gradual but highly significant loss of transparency to the public as well as local, state, and federal officials of government. In dealing with the public and gaining their support, it is important to keep things simple and understandable. Because transparency is a concept that the public understands and is easily associated with water quality, Secchi depth has been a very effective way of communicating the lake problems, their causes, and their most probable solutions.

In California a well-known saying, "Keep Tahoe Blue", was made a common household slogan by a conservation organization known as the League to Save Lake Tahoe. Bumper stickers bearing the phrase appear on thousands of cars all over the state and the League boasts both a large membership of 4500 and considerable influence in the northern parts of the states of California and Nevada. The slogan has been particularly effective in combination with Secchi measurements in directing both public attention and concern to the early stages of eutrophication occurring in the lake. Examination of the long-term Lake Tahoe Secchi record indicates an average loss of transparency of 0.33 meters per year over the last 40 years, with considerable interannual variability (Figure 2). The regression of this data is significant at the 0.001 level, indicating that there is less than one chance in 1000 that this trend is due to chance alone. If, however, one had taken only the period between 1983 and 1987, they would have come to exactly the opposite conclusion that would be grossly in error in predicting the fate of the lake, because transparency increased each year during that period. In fact, someone unschooled in the importance of looking at long term data in ecological questions, took this short run of data, out of context from the long term record, to Washington, D.C., the nation's Capital, and claimed that the problems at Tahoe had been solved and further government investment in lake restoration was unnecessary. It is obvious from viewing this long-term data set that it must be considered in its entirety rather than looking at short portions of data.

The explanation for this short-term reversal of the significant trend of declining transparency was drought conditions that prevailed during this period. These conditions reduced nutrient and sediment inflow to the lake, while at the same time, the lack of late-winter storms resulted in a shallower than normal depth of mixing during the fall and winter mixing periods. Because of this lack of deep mixing, significant portions of the nutrients and fine sediment particles, which had not yet been deposited but remained trapped in the depths of the lake, were not returned to the lighted (euphotic) zone during shallow mixing years and thus did not promote algal growth and reduce transparency. Over the four decades now covered by our monitoring program, algal growth rate as

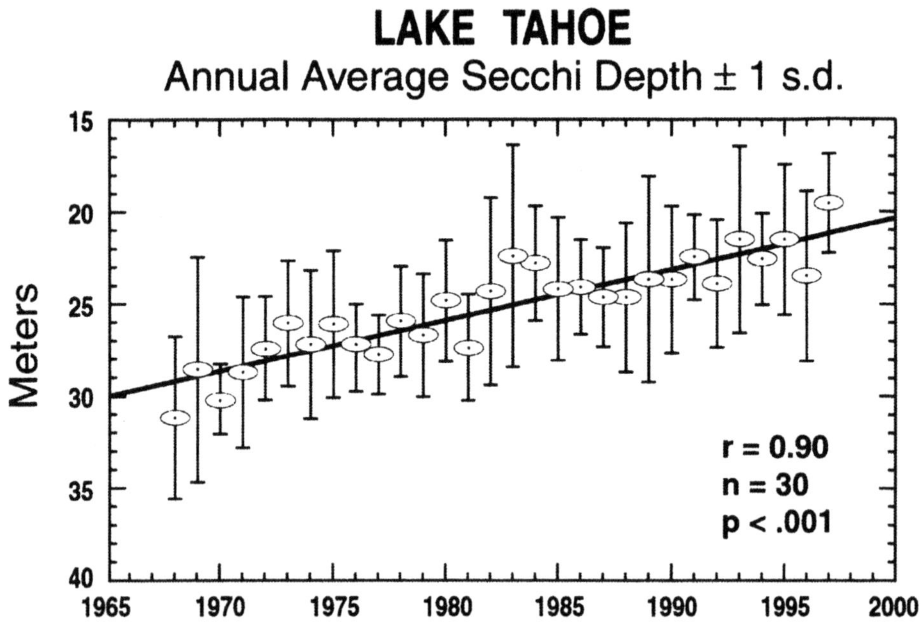

Fig. 2. Lake Tahoe Secchi disk measurements with one standard deviation from the mean indicated. Each year is the average of 35 measurements selected for optimum viewing conditions. The data indicate a progressive loss of transparency of about 0.33 meters per year.

measured *in situ* with the sensitive 14-carbon method (Goldman 1963) has increased at about 5% per year (Figure 3).

In Asia, Lake Biwa, which supplies water for 14 million people, could certainly benefit from an agency mandated by the Japanese government to serve as an umbrella to coordinate anti-pollution measures in the extensively developed and farmed watershed. This should make it possible to take more effective and unified actions to reduce nutrient loading that increases both cyanobacterial growth and the cost of water treatment. This condition threatens the future use of an incredibly important water supply. Another well-known and important Asian freshwater lake is Lake Baikal in Russian Siberia, the world's deepest and oldest lake. It actually contains 20 percent of the entire world's unfrozen surface waters as well as over 2500 endemic species of plants and animals. It is probably Russia's single most valuable natural resource. Although the watershed has a very small population, industrial pollution is significant, largely attributed to small industry at Ulan Ude on the Selenga River, the main tributary to the lake, and a paper mill on the lakeshore. Fortunately, the dilution by the enormous

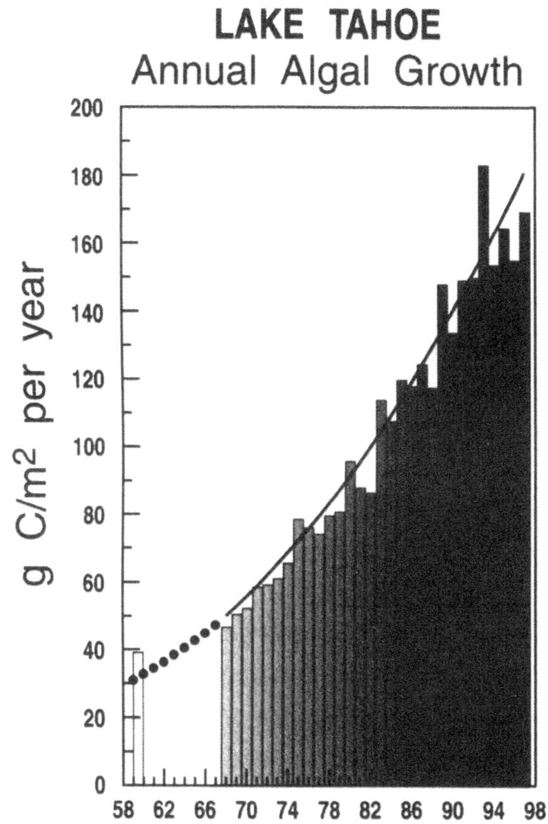

Fig. 3. Primary productivity measurements by the 14-carbon method. Each year is represented by the average of 14 *in situ* measurements to a depth of 115 meters taken at approximately two-week intervals throughout the year. There has been a progressive increase in algal growth rate of approximately 5% per year, with interannual variability strongly correlated with the annual depth of mixing and nutrient inflow from tributary streams.

volume of the Southern basin is great and the main sources of pollution noted above are near the Angara River, which is the single outflow from the lake. Far from the central government in Moscow, Baikal has long been the focal point of environmentalism in Russia. Although the pollution attributed to the Baikalsk paper mill has been at the center of the controversy, the plant has remained in production for over four decades. Generally considered to be incompatible with tourist and fishery use, resource discussions for conversion of the plant to a less polluting activity have been underway for years. Since the lake has been given new status on the global stage by being designated as a World Heritage Site, water pollution from the many watershed inputs is reported to have been reduced, but air pollution from the paper mill still remains a terrestrial and aquatic problem that has received little attention.

Developing Political Awareness for Managing the World Water Crisis

One approach that has been particularly important and successful at Lake Tahoe in raising political awareness has been to take high-level political figures and their staff out on the research boat to view lake conditions firsthand. The 1997 visits of President William Jefferson Clinton and Vice President Al Gore for the first Lake Tahoe Summit meeting is a particularly noteworthy example since before that time no political figures of their stature had made environmental concerns so newsworthy. Spending a full hour aboard the UC Davis Tahoe Research Group research vessel *John Le Conte* (Figure 4), they had the time to be fully briefed on conditions in the lake and the most likely future of the lake if serious remedial efforts were not instigated. Further, political figures derive substantial benefit from public exposure related to environmental problem-solving. The President's and Vice President's historic visit to the lake was followed by similar visits by Senators, Congressmen and the governors of the states of California and Nevada. In 2001, another group of officials, including the directors of several Federal agencies from President Bush's new administration, participated in a second summit meeting at the lake and also went out on the research vessel for briefing on the lake's condition. A third summit directed by California's famously influential Senator Dianne Feinstein was held in October 2002.

It is important that local, national, and world leaders first recognize and then face in an effective manner the problems associated with the decline in both quantity and quality of our limited surface waters. It is the urgent responsibility of scientists everywhere to communicate effectively with the political leadership and provide adaptive management recommendations in clear, unambiguous language. For some unexplained reason, the essential nature of this commodity for the survival of the plant and animal inhabitants of the planet has been lost in our headlong rush to produce more and more people, food, fiber and machines from our increasingly limited natural resources. The fact that water remains an under-valued commodity is key to the understanding and future management of this essential natural resource.

Air pollution has become a major contributor to the pollution of water supplies worldwide and clearly illustrates the need for a more global perspective in managing water resources essential to this and future generations. The time has come when it is

photo courtesy of *The Sacramento Bee*

Fig. 4. U.S. President Clinton and Vice President Gore learn firsthand from Charles Goldman about Lake Tahoe's water quality aboard the University of California's Research Vessel *John Le Conte* in July 1997.

probably necessary to create a global surface and perhaps even groundwater monitoring agency, either as an extension of the United Nations UNEP program or through ILEC, to extend our local and national activities to a coordinated global approach for dealing with the serious problem of declining usable water supplies. The upcoming World

Water Conference in Japan in 2003 may well be the place to start this initiative by coordinating monitoring and information exchange between the nations of the world.

References

BYRON, E. R., AND C. R. GOLDMAN. 1990. The potential effects of global warming on the primary productivity of a subalpine lake. Water Resources Bull. **26:** 983-989.

EDMONDSON, W.T. 1991. The Uses of Ecology: Lake Washington and Beyond. Univ. Washington Press. 329 p.

GOLDMAN, C. R. 1963. The measurement of primary productivity and limiting factors in freshwater with Carbon-14, p. 103-113. *In* M. S. Doty [ed.], Proc. Conf. on Primary Productivity Measurement, Marine and Freshwater. U.S. Atomic Energy Commission Report No. TID-7633.

_________. 1978. Ecological aspects of iceberg transport from Antarctic waters, p. 642-651. *In* A. A. Husseiny [ed.], Iceberg Utilization. Proc., First Internat. Conf., Ames, IA.

_________. 1994. The Sea of Aral, p. 137. *In* M. Seely [ed.], Deserts. Weldon Owen Pty Limited, McMahons Point, NSW, Australia.

_________. 2000a. Baldi Lecture. Four decades of change in two subalpine lakes. Verh. Internat. Verein. Limnol. **27:** 7-26.

_________. 2000b. Management-driven limnological research. Arch. Hydrobiol. Spec. Issues Advanc. Limnol. **55:** 257-269.

_________, AND E. DE AMEZAGA. 1984. Primary productivity and precipitation at Castle Lake and Lake Tahoe during twenty-four years. Verh. Internat. Verein. Limnol. **22:** 591-599.

6-2. Lake Management Requirements from a Local Perspective

M. R. James[1], W.Vant[2] and C.Severne[3]

[1] *National Institute of Water and Atmospheric Research Ltd, P.O.Box 11-115, Hamilton, New Zealand.*

[2] *Environment Waikato, P.O. Box 4010, Hamilton*

[3] *National Institute of Water and Atmospheric Research Ltd, P.O.Box 8602, Christchurch, New Zealand.*

Abstract

The aim of lake management is to satisfy demands of lake users in a way that is compatible with protecting and enhancing long-term characteristics and values of the lake while minimizing conflicts. There is an urgent need to protect lake resources and their catchments to sustain environmental, social, cultural and economic values and growth. Protection and sound management is essential for local communities and particularly indigenous communities who rely on these highly valued resources. Clearly there are a wide range of uses, which need to be considered with lake management. The value placed on these uses depends on the specific circumstances of a lake and thus must be looked at from a local perspective. This needs to involve local communities, government, scientists, indigenous groups, business interests, end-users and interest groups. In this chapter we address the issue of lake management from a local perspective using Lake Taupo, New Zealand, a large oligotrophic lake, as a case study to demonstrate the principals. Policies are being put in place to protect Lake Taupo from further degradation based largely on community concerns. Legislation is often essential and can be best achieved in terms of regulatory frameworks at the local or regional level.

Introduction

"There are global issues with lake management but the real action happens locally" (Margaret Catley-Carlson, Chair, Global Water Partnership).

Water issues can be considered from generic global issues down to the local community scale. We are all well aware of the need to conserve high quality water, with an estimated 1.1 billion people having no access to safe drinking water and clean water is fast becoming a global commodity. Lakes are one of the major sources of this water but are also highly valued for a range of other, often competing uses, yet there are rarely restrictions on access or its use by the public. As highlighted by Korth and Klessig (1990), lakes are prime candidates for the "tragedy of the commons" where a resource is used by everybody but managed by no-one. Although there have been significant advances over the last decade, the potential for such tragedy still exists for many lakes as they come under increasing pressure for water supply, recreation, waste disposal, fisheries, cultural and spiritual needs. Many lakes have been heavily exploited and are considered a free and limitless resource.

The aim of lake management is to satisfy demands of lake users in a way that is compatible with the protection and enhancement of long-term characteristics and values of the lake, while minimising conflicts. A key issue world-wide is the increasing degradation of these highly valued natural resources and there is an urgent need to protect these resources and their catchments to sustain environmental, social, cultural, recreational and economic values and growth. This protection is particularly essential for local communities (the people who are intimately linked with lakes through living, working or own property in the catchment) who rely on these resources in their daily lives. It has been shown in many cases that those that live with the consequences of any actions are the ones best placed to address and balance the competing challenges and opportunities. It is important however, that we learn from others' mistakes and information and experiences are shared between communities and countries. What then is required for lake management from a local perspective, if this is where effective management is ultimately driven?

Effective lake management is often a joint effort between central and local government, the scientific community, multiple lake interest groups, end-users and industries but there is also a general consensus now that it is only the care and concern of local communities that can overcome the potential for the "tragedy of the commons" syndrome. Central, state and regional government can provide some enforcement of activities, like boating and fisheries, and can provide the framework for lake management but often cannot provide effective management at the local level. The State of Wisconsin in the USA is a classic case. There are over 15,000 lakes in the state, and for many of these, management relies heavily on the care, concern and involvement of the local communities. Wisconsin was also the first state to put in place controls along shorelines which are administered by local counties.

In this chapter we address the requirements for lake management from a local perspective addressing the major principals involved. These principals include the need for good integrated management plans which identify the values, issues and goals and suggest or identify options and potential solutions. We then highlight the involvement of different sections of the community and government. Finally we use Lake Taupo in the central North Island, New Zealand as a case study to demonstrate how these principals might be applied. The New Zealand case has a bicultural dimension and provides a New Zealand and international model to address lake management issues. Ideally there should not be a need for lake management and too often this involves restoration of a degraded water body because issues are not addressed or managed early enough. While deterioration of lake water quality can in many cases be rapid, restoration can take many years and is generally considerably more expensive than putting in place appropriate management plans. The aim in the Taupo case is to prevent serious deterioration in water quality before it is too late.

Local Lake Management Plans and Programmes

Each lake and its management is unique because of its physical setting, catchment use, lake use and the social, cultural, political and economic statutes which operate in the region. Much however, can be learnt from experiences of lake managers and scientists who have encountered similar problems. Central and local governments are generally held responsible for solving management problems but this rarely works unless local communities are involved right from the start in having input to management plans.

Appropriate management plans are necessary as a framework and can be issue-specific but in many cases they address a range of management issues and options depending on the local circumstances. While some management plans may have legislative backing most, such as those implemented in some Kenyan lakes (e.g., Lake Naivasha and Lake Nakura, Abiya 2001), are voluntary and rely on public awareness of the need for proactive planning. In the case of Lake Victoria an Environmental Management Project (Mwiva 2001) has been set up to formulate management plans but these will only be sustainable through partnerships among all those involved in the lake and its catchment, along with government and non-government organisations.

The key to these plans is to develop mechanisms to allow collaborative management by local stake-holders and communities with guidance from scientists, and guidance and funding from local and central government. The most appropriate scale for these plans is the watershed which allows for integrated management of catchments and the receiving water bodies. It is essential, for example, that people living in the upper part of catchments appreciate the effects of their activities on downstream communities and in particular the lake ecosystems and communities who live around them. There can however, be real difficulties with participation and different views on issues and approaches depending on whether these are viewed from the local, national or global level. As discussed later in this chapter there has been a significant move towards decentralisation and devolvement of responsibilities for management plans to local government.

The Lake Champlain Basin Programme (LCBP) is a model partnership programme for watershed management (Howland 2001, Watzin 2001). Part of the reason for the success of this programme is the level of funding where some US$12M of federal funds have been provided for plan implementation. In many other cases, unfortunately, the management plans have faltered because of a lack of appropriate planning and funding. In the Lake Champlain case six advisory committees were set up to ensure involvement by a range of government and non-government bodies. These include: a Technical Advisory Committee involving researchers, technical experts, representatives from agriculture, planning and economics; Education and Outreach Committee; a recently formed Cultural Heritage and Recreation Advisory Committee; and a Citizen Advisory Committee, in each of the 3 states. These ensure there are good linkages between the local communities with the LCBP when developing management plans which can be actioned in their own communities and catchments, with the necessary funding provided through federal, state and provincial agencies.

In the case of the Philippines a tripartite group, the Philippine Watershed Management Coalition, was set up in 1998, involving the Department of Environmental and Natural Resources (DENR), Local Government Units (LGUs) and Non-Government Organisations (NGOs). The main role of this tripartite group is education advocacy and capacity building to ensure there is appropriate planning and management of watersheds (Gomez 2001). This is another good case where there is an excellent framework set up for local community involvement in lake management plans but relies heavily on central government for funding.

Key Requirements for Lake Management

Increasing pressure from population and agricultural growth and intensification mean there is an urgent need for an integrated planning and management approach involving catchments and the associated lakes in most countries. Generally there are five key features which need to be considered when developing management plans. Firstly there must be a good description of the natural resources, the issues and community values associated with the lake and its catchment must be identified through wide consultation with those involved in the lake. Then goals, options and strategies to meet those goals must be clearly set out. In this section we discuss these requirements in more detail and in a later section we demonstrate the application of these principals in the case study with Lake Taupo.

1. Description of the natural resource

   Without scientific input providing a very good description of the natural resources and an understanding of the physical, chemical and biological processes involved it would be very difficult to put in place appropriate management strategies. A good example is some of the lakes in the central North Island of New Zealand. Unlike many northern hemisphere lakes where phytoplankton growth is generally phosphorus limited these lakes have been shown to be nitrogen limited for at least part of the year. If management strategies were put in place for reducing phosphorus loadings and nothing was done about nitrogen loadings then management strategies would have been futile. Likewise a description of rare and endangered plant and animal species and key habitats is essential for maintenance of biodiversity, requiring specific management strategies for the lake. Some lakes developed for hydroelectric purposes on the other hand may have very little value for maintaining biodiversity. These cases highlight the importance of having a good understanding of the natural resources, ecosystem functioning and associated values for the local community before starting to formulate management plans.

2. Identifying the issues and community values

Clearly there are a wide range of uses which need to be considered with lake management. These include Agriculture, Forestry, Ecosystem and Habitat Preservation, Biodiversity, Recreation, Commercial Fisheries, Scenic Values, Tourism, Hydroelectric Storage, Waste Disposal, Water Supply and Flood Control.

Identifying the issues and values associated with a particular lake is essential before setting up management strategies. Issues can often be conflicting and range from national problems, such as supplies of potable water for large regions, to the local level where communities may be concerned about issues such as eutrophication, habitat for endangered species, and a commercial or recreational fishery. Community values in particular will vary between lakes and catchments and thus values must be looked at in a local perspective. In New Zealand, for example, lakes such as Otamangakau in the Central North Island and some of the Waitaki lakes in the South Island have been formed through damming of rivers for hydroelectric purposes but have since also been extensively developed for recreational fisheries. Lake Otamangakau has become renowned for its trophy size salmonid fisheries. Lake Waikaremoana is part of a National Park in New Zealand, highly valued for its cultural, ecological, intrinsic, and recreational values but is also the main storage lake for 3 power stations downstream. Many lakes in Norway have been developed expressly for hydroelectric storage with lake level fluctuations in some lakes of over 100 m depending on demand for electricity.

Some large New Zealand lakes like Taupo (see case study later in this chapter) have few inhabitants in their catchment and little urban development but the shoreline catchments are used extensively for agriculture and forestry. Urban pressure on large lakes in other countries is often considerable with over 1.5 million people in the Lake Biwa catchment in Japan and 20 million in the catchment of Lake Victoria which is bordered by 3 countries Kenya, Uganda and Tanzania.

A primary use for many lakes is water supply. Lake Biwa in Japan, for example, supplies drinking water for 14 million people in the Kinki region but its waters are also extensively used for agriculture, and the lake is highly valued for its commercial and recreational fisheries, tourism, is one of the ancient lakes of the world with considerable scientific interest and the lake has significant religious and cultural values. Another example is the Great Lakes, encompassing the group of large lakes on the border of Canada and the USA. These lakes have 40 million people in their catchments who rely on these resources for drinking water along with water supply for industry, agriculture, tourism, recreation and commercial fishing.

3. Setting achieveable goals

Once the issues and values have been identified then management goals or targets need to be set to maintain or enhance those values. These must be achievable and affordable if they are to succeed, bearing in mind all the different sectors that will be affected. Establishing what is an acceptable compromise needs significant input from all the local community as generally this is the section of society that will be most affected by changes in catchment practices and the lake's condition. In the case of water quality the greater the controls required the higher the costs to groups such as farmers, industry and other end-users who may have their activities curtailed. It is also essential that goals are consistent with regional policy statements and plans. An example of specific goals for a lake will be demonstrated in the case study on Lake Taupo.

4. Options and solutions or strategies

Once issues and goals have been identified and documented, options and strategies can be explored. Putting options, based on sound scientific and technical advice, to the various stakeholders and the local community early on in the process provides them with the opportunity to voice concerns and more importantly utilise their local knowledge to openly discuss the benefits and disadvantages from environmental, cultural and socio-economic perspectives.

Solutions or strategies to achieve the goals must be realistic and take into account conflicting end-users and other values and goals of lake management. For example, the solution to overfishing may be to limit catches but if the local community rely on this as a source of food and the lake is important for recreational fishing and tourism then consideration should be given to releasing juveniles and enhancing spawning habitat as an option. Similarly reducing nutrient loadings from lake catchments should involve an assessment of improved and more efficient technologies rather than directly limiting the production of beef or dairy herds. The key consideration should be the effects on the water courses and what mitigation strategies can be put in place.

5. Consequences

The consequences of any strategies must be carefully weighed up. Many attempts to reduce nutrient loadings to lakes have not considered the socio-economic effects on the local communities, such as agriculture and forestry, which may have heavy controls placed on them. Financial consequences could mean they go out of business or move off the land thus affecting local infrastructure and regional economies. Some form of cost analysis and documenting the benefits and disadvantages is often needed to put conflicts into perspective but these are rarely considered as part of lake management draft plans. Assessing the consequences may also involve some form of consensus on what is acceptable to the community. For example, what is the level of activity that should be allowed before an unacceptable level of change in environmental indicators for a particular lake is reached. This type of analysis should be carried out by representatives of the local community including indigenous groups, stake-holders, industry as well as scientific and technical experts. This approach, commonly referred to as "Limits of Acceptable Change", also gives the community an indication of the scale of impacts to expect and provides a participatory decision-making process. Such an approach was successfully used recently where dredge spoil was to be disposed off the Queensland coast in Australia (Oliver 1995).

Identifying issues, setting goals and developing options and solutions should be done in partnership between government and the local community. Depending on the specific lake and the issue this may involve the local community, scientists, local government, indigenous groups, end-users and stakeholders, NGOs, local trusts and committees (for worked examples see case study later in this chapter). In the next section we outline the involvement of these different groups in lake management with an emphasis on the local perspective.

Local community

It is now widely recognized and accepted that local communities must have a much greater involvement in decision making and implementation of management plans for lakes. Experience shows that those who are most closely connected and affected by any consequences of management are best able to address and balance the conflicting challenges and opportunities. Involvement in meetings, workshops, advisory committees and strong local partnerships are essential for effective management ensuring a sense of ownership and making the most of collective local wisdom and experience. An excellent way to involve the community is through advisory groups such as the Education and Outreach and Citizen Advisory Committees set up for Lake Champlain (see Howland 2001, Watzin 2001).

Indigenous groups

Indigenous groups have a vast store of knowledge passed down through generations and have lived in harmony with lakes for time memorial. Water has always played a major part in their spiritual and cultural values and these values must be incorporated in any planning and decision-making. Many groups are starting to be more pro-active and play a very important role in lake management with some now developing their own environmental strategies which must be incorporated from the beginning in any strategic framework. Indigenous groups already play a major role in this process in places like Indonesia and Kenya where their experience and understanding of environmental conservation have been invaluable. Another example is the indigenous groups in New Zealand (see case study) and Canada who are now heavily involved in decision making on lake management. The framework for lake management must provide a process for their involvement which is acceptable and consistent with their cultures.

Traditional wisdom and religious practices have also been incorporated as part of some management plans and in decision making. This is particularly important where there are sacred areas or religious consideration such as Lake Toba in Indonesia where there are certain behavioural restrictions around the shores of the lake (Hehanussa 2001)

End-users and stake holders

Agriculture and forestry are major stakeholders in many lake catchments and shoreline developments can potentially have significant environmental impacts if not managed and controlled properly. They are also an important part of the community, providing jobs and income for the regions.

Farmers have interests in maximising production but in a sustainable way taking into account land-use, the waterways draining the catchments and lake environments. They are particularly concerned with economics of their operation but equally are concerned with improving efficiencies. A major challenge is to provide an appropriate mechanism for their involvement and to convince farming groups of the benefits of sustainable use of waterways, such as lakes, and more efficient ways to operate and minimise impacts on the natural aquatic environment. Their buy-in to management plans and solutions is essential but in many cases it does require a change in mindset. Farmers are playing an increasingly important role in partnerships involved in lake management such as the Lake Victoria Environmental Management Project (Mwiva 2001). Reducing the quantity of fertiliser used, recycling of agricultural wastes and improved effluent treatment and improved technologies for the livestock management, such as dairying, are some of the considerations to be taken into account and are a critical part of management plans for lakes like Taupo in New Zealand and Biwa in Japan.

Tourist operators often rely on a clean, green image or natural beauty as the basis for their industry and thus have a vested interest in maintaining the quality of lakes. Aesthetic values and recreational fishing can be major drawcards for visitors to lakes and their catchments. The interests of recreational groups must be considered as they can be significantly impacted by declining fish populations, whether it is from increased pollution or invasion by exotic species.

NGOs, Local trusts and committees

Non-government organisations play a crucial role in development and implementation of lake management plans. They often provide a critical link between the local community and central or state government and can be excellent advocates for effective and proper lake management. Some NGOs are set up with a wide range of interest groups and can provide an umbrella organisation representing local and regional groups. An example is "Sinchan Sahyog" in India which is an NGO consisting of farmers, scientists and engineers who all actively participate in development of plans, decision making and plan implementation. Such groups can be particularly effective in tackling difficult problems associated with implementation and co-ordination for lake management at a local level.

In some cases management plans aim to prevent degradation of the natural resources of lakes but in many cases management aims to restore lake values. Restoration measures put in place to restore Lake Victoria have included improved catchment practices, comprehensive management plans for fisheries involving local fisherpersons, community participation and incorporation of indigenous knowledge, opinions and views, legislation brought together under one umbrella and setting of regional water quality standards and pollution control measures (Mwiva 2001).

Local and central government involvement

A significant development worldwide in lake management has been the devolvement of responsibilities for lake management and decentralisation from central and state government to local government. There are many examples of this. For example, in the Philippines the management of small watersheds (up to 5000 ha) was devolved from the Department of Environment and Natural Resources (DENR) to Local Government Units (LGUs) in 1991. While this development has greatly improved the involvement of local communities in collaborative management of lakes, and is the most appropriate level for involvement by government, too often it has not been accompanied by the provision of appropriate funding, development of expertise and legislative provisions to enable them to raise revenue for lake management projects. Training sessions to ensure middle managers with LGUs in the Philippines had the appropriate level of skills has been important and successful but like most situations there have often been problems with funding. There is a danger however, that communities will still rely heavily on government bodies to fund and manage these resources and a change in attitude to more of a partnership relationship may need to be encouraged.

Donahue (1996) summarised the development of water resource management in the Great Lakes bordering Canada and the United States in a memorial lecture and traced the gradual devolvement of responsibility to the local and regional level. Five eras were identified which could equally apply to other countries but on different time scales. These eras were:

- Resource Development Era through the middle of the 19th century when the emphasis and lake management initiatives were focussed on development such as provisions for water transport.
- The Transitional Era in the latter half of the 19th century when water resource development became the responsibility of multi-jurisdiction institutes with multiple objectives. It also marked the setting up of various commissions and the notion of watershed management and drainage basins was first introduced.
- The Federal Leadership Era took us through to the middle of the 20th century and marked the beginning of federal domination of legislation and water management institutions. An example is the Inland Waterways Commission which emphasised the need for comprehensive planning, co-operation between governmental, private and public sector groups and a formal institutional structure at the federal agency level.
- The River Basin Era that continued through to the mid 1980s was characterised by emerging partnerships, institutional development at the river basin level and for

the first time an emphasis on environmental protection and resource management signalling a move away from development as the major objective.

- Finally we are now in what Donahue (1996) termed the New Era where top down control by government has been devolved in favour of a bottom up approach characterised by partnerships and increased citizen involvement. This has created greater awareness and ownership at the local and regional level, stressing voluntary compliance and input to decision making and lake management plans by the people most affected and closest to the issues.

Similar trends can be seen, perhaps on different time scales, in many other countries. Unfortunately this latest era of increased devolvement also places the economic burden on local government and communities who may not have the technical resources. In most cases there has been no increase in state funding. The latest plan for the Great Lakes, the "Great Lakes Strategy 2002 – A Plan for the New Millenium" aims to address major issues but there has been no additional funding from governments. However, the plan demonstrates a greater commitment to establishing partnerships between federal, state and tribal agencies.

It is generally agreed that local government is the right level of involvement but they must be empowered with the right legislation and authority. In many countries there are a number of acts governing responsibilities for lake management. Bringing these together under one legislative act, as was done in 1999 for Lake Victoria under the Environmental Management and Co-ordination Act, has seen a major improvement in integrated approaches at the local level, helping to protect and conserve natural resources such as lakes. In some cases legislation has been enacted to ensure end-user involvement at the local level. In 1992, for example, Mexico introduced participation of water users and interested third parties as mandatory under water laws (Kurtycz 2001).

The Environmental Management and Co-ordination Act was introduced for Lake Victoria in 1999 bringing together a number of legislative acts under one umbrella with the overriding objective of protecting and conserving natural resources. Regional water quality standards are also to be put in place with laws for pollution control. A number of other governments have also moved to bring a number of environmental protection acts under one umbrella and an example of this is included in the following case study.

Case Study

In the previous section we outlined the general principals for approaching lake management, from setting up a lake management plan to identifying options and solutions and the need for involvement of different sections of the community and government. In this section we describe and work through the application of these principals in a case study.

Lake Taupo, New Zealand

Lake Taupo on the central volcanic plateau, North Island, New Zealand is a large oligotrophic lake with clear water and high water quality. The lake is highly regarded for its cultural and spiritual values by the indigenous people, and for its ecological, social and recreational values (see Plate 1). Taupo, is a small town with a resident population less than 20,000 but is one of New Zealand's most important visitor destinations, with direct visitor spending of about US$60 million per year much of this due to the world renowned rainbow trout fishery. With downstream spending, this approaches a total of nearly US$100 million per year (Anon 2002).

A community consultation process undertaken in 1998 identified fourteen goals for the lake and its catchment, ranging from "clear water", "safe swimming" and "safe drinking water" to "outstanding scenery", "recreational opportunities" and "cultural values" (LWAG 1999). In 2001, central government and various other agencies began work on a project—"2020 Taupo-nui-a-Tia"—to develop a long-term action plan to protect the health of the lake and its surrounds taking into account the fourteen community goals. This involved extensive consultation with groups including landowners, Maori trusts, District Council staff farmers, dairy companies, forest managers and tourist operators.

Lake management plans and programmes

The 2020 Taupo-nui-a-Tia Project provides a long term vision for Lake Taupo and its catchment integrating social, cultural, environmental and economic knowledge. The project is funded and supported by central and local government, Tuwharetoa Maori Trust Board and the Lakes and Waterways Action Group (a local interest group). A major focus of the project is communicating and involving the community in helping to develop actions and solutions. There are three major interactive strands – iwi or the local indigenous people, socio-economic and science. All are working towards a joint Management Strategy for Lake Taupo and providing input to local and regional plans as part of the legislative framework.

Plate 1. Images from Lake Taupo showing the important cultural, scenic and recreational values associated with this large oligotrophic lake in the central North Island, New Zealand.

The first phase for the science strand involved reviewing current knowledge, threats and knowledge gaps in relation to the 14 values mentioned above. This phase is now complete and along with the other strands the focus has shifted to comparative risk assessment for each value, to be followed by scoping and assessment of risk reduction options. While the 2020 Project involves long term visions and planning, closely aligned to this project is the Lake Taupo Water Quality Project which is the first active step to help address one of the key issues – declining water quality.

Nutrient inputs from intensified catchment use has been identified through the 2020 Project and Lake Taupo Water Quality Project (LTWQ) as a major threat to the high water quality of Lake Taupo. The Waikato Regional Council's aim under the LTWQ Project is to protect lake water quality by managing land use in the catchment under section 30 of the Resource Management Act (RMA -see below) ensuring that it is consistent with a number of the community goals.

To address concerns raised by various groups and enhance the scientific basis for management, several studies have been initiated (e.g. Elliott and Stroud 2001, Hadfield et al. 2001, Spigel et al. 2001) with the aim of enhancing the information available to underpin decisions on managing the catchment and ultimately Lake Taupo's water quality. At the same time, it is recognised that it is not realistic to expect to have a perfect technical understanding of this complex ecosystem. This means the lake must be managed based on the best information available at the time and scientists must play a key role in this process.

The key issue being addressed in the Lake Taupo Water Quality Project is catchment land use and impacts on the lake. The approach being taken follows the principals of what is required for lake management in terms of the issue, management goals and targets, options and solutions which were discussed in the first section.

1. *Issue*: *Intensifying catchment land use and declining lake water quality.*

   Changes in water clarity in Lake Taupo are mainly determined by the presence of dissolved organic and particulate matter with maximum algal productivity occurring in late winter. The upper water column is generally clearest during late spring and summer when a lack of nutrients in the surface waters limits algal growth. Studies over the past 20 years have shown that this nutrient is nitrogen for much of the year (White and Payne 1977, White et al. 1986, Hall et al. 2002).

   Water clarity (secchi disk depth) measurements up until the late 1980s averaged 15.3 m but since 1990 have reduced to 14.2 m and summer maximum transparency has decreased from 20 m to 16 m. This decline is likely to be the result of increased nutrient loadings from catchment use combined with natural processes such as large scale climatic changes. Proposals to intensify pastoral farming have highlighted the need for a better understanding of the links between catchment use, land-use changes and in-lake processes.

Increasing loads of nutrients to lakes generally result from the intensification of human activity in their catchments. Both point sources of nutrients, especially from industrial and municipal sewage discharges, and non-point or diffuse loads from catchment land use can be important. In New Zealand, diffuse sources of nutrients have been shown to be generally more important than point sources (e.g. Rutherford et al. 1987). Whereas the restoration of 23 eutrophic lakes in Europe and North America primarily involved wastewater diversion or treatment (Uttormark and Hutchins 1980), only two New Zealand lakes—Horowhenua and Rotorua—have been shown to be markedly affected by such point sources. In both cases, diversion of sewage effluent away from the lake has been undertaken (Anon. 1990, see also Vant and Gilliland 1991; BOPCC 1975, see also Rutherford et al. 1989). Substantial reductions in nutrient loads to other degraded lakes in New Zealand will involve the considerably more difficult management of diffuse sources of nutrients (e.g. Vant 2001).

The nature of the land tenure in the Lake Taupo catchment, and the low fertility of the volcanic soils meant that prior to the 1950s there was little development of the catchment for agriculture. From about 1970, government and private developers developed land in the northern and western parts of the catchment. By 1973 about 17% of the catchment was in pasture, increasing to its present 22% by 1996. The magnitude of the nutrient loads entering the lake from the various parts of the catchment was well-documented in the 1970s (Schouten et al. 1981) and has been the focus of various modelling studies since then (e.g. Elliott and Stroud 2001). However, it has recently become apparent that there has been a substantial delay between the period of catchment development and the appearance of increasing nitrogen concentrations in streams draining pasture areas (Vant 2001). In addition, a preliminary study to determine the age of the water in one stream draining an area of pasture has shown that about two-thirds of the water in this stream at baseflow is more than 35 years old—that is, that it is unlikely to have been affected by the recent land use changes (Morgenstern 2001). As time goes on, and the water in this stream is progressively replaced by newer water that has been affected by agriculture, we can expect that nitrogen concentrations will continue to rise.

Vant and Huser (2000) assessed the expected effect of any further intensification of agriculture in the catchment on lake water clarity. They concluded that a major intensification could increase algal biomass in the lake by 20–60%, causing water clarity to decline by 20–40%. Spigel et al. (2001) subsequently refined this analysis, using a substantially more detailed modelling approach. They predicted from a preliminary model that a major intensification of agriculture in the catchment would cause algal biomass in the lake to increase by 47%, resulting in a 2.5 m reduction in Secchi disc depth. Furthermore, they concluded that even with no further intensification of land use, Secchi depth will continue to decrease, resulting in a further decline of 1.1 m by 2020 due to predicted changes in climate.

2. *Management Goals and targets*

The major goal identified by the community, indigenous groups, stakeholders and lake managers is to have pristine clear water, which is safe to drink, has high water clarity and low amounts of weed and slimes around the edges.

As part of the Lake Taupo Water Quality Project the local Regional Council have worked alongside the various groups to identify realistic targets, goals and options (Figure 1). These are:

Option 1 – Better water quality through reduction in nutrients coming from catchments with less intensive farming to bring it back to pre-development conditions.

Option 2 – Maintain present water quality which would involve some reduction in nutrients coming from the catchments because of lag times and with no further intensification.

Option 3 – Accept slightly lower water quality but maintain the status quo with land-use and no further intensification

Option 4 – Accept lower water quality i.e no controls on land use and allowing intensification to continue to increase.

Option 1 was clearly not favoured as it would ultimately result in no farming in the catchments and Option 4 was not favoured by interest or stakeholder groups. The overall aim then is Option 3 in the short term but to aim for Option 2 of reducing nutrient loads in the longer term. The approach taken with Lake Taupo is similar to that set out in the Mother Lake 21 plan for Lake Biwa in Japan (Organising Committee of the 9th International Conference on the Conservation and Management of Lakes, 2001). This sets out goals and timeframes to ensure sustainable development and conservation of natural

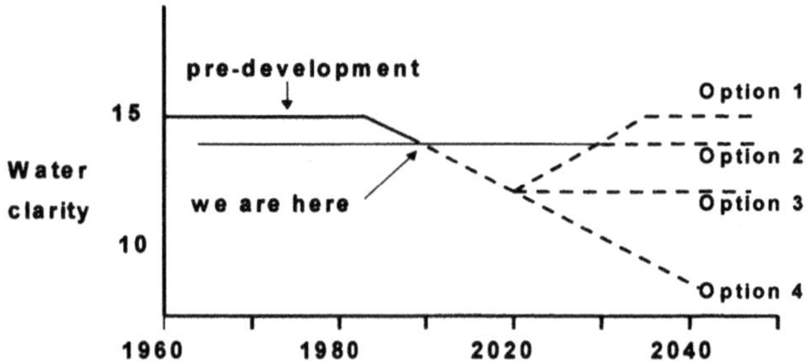

Fig. 1. Schematic diagram showing changes in water clarity with time under the different options listed above.

resources for future generations. The goal for Lake Biwa is that by the end of 2020 the Lake will be in a state suitable for future generations with a vision of further improvement by 2050. In terms of water quality the objectives are to improve water quality by restoring loadings to those of the late 60s and in the second stage to return water quality to levels before red tides and algal blooms developed.

3. *Options and solutions*

Further studies on groundwater storage in the Lake Taupo catchment, and thus the delay between land development and increasing loads of nutrients in the inflowing streams, are currently underway. In the meantime, a provisional estimate has been made that the nitrogen load from developed areas of the catchment at steady-state is likely to be up to about 20 (± 10) percent greater than at present. This means that maintaining the current nitrogen load to the lake—and thus maintaining current lake water quality—is likely to require a de-intensification of land use in parts of the catchment. As a result, a provisional target has been identified of reducing the steady-state nitrogen load from pasture and urban areas by 20%.

At this stage, both regulatory (i.e., RMA-based) and non-regulatory methods for achieving the reduction in nutrient loads to Lake Taupo are being considered. Land use restrictions would be likely to see a reduction in intensive farming, and a shift to lower nitrogen-yielding activities (e.g. forestry). Both land values and incomes would be predicted to fall as a result and must be taken into consideration. At this stage the non-regulatory methods being considered are focussing on the possibility of multi-party financial assistance to affected landowners to allow them to cease farming in the catchment or to shift to lower-yielding activities, perhaps involving service agreements with the regulatory authorities. Options currently being discussed with the community—both the directly affected landowners and the wider community—include:

- Voluntary change involving education and encouragement - this is likely to only have limited success.
- Financial incentives through assistance with fencing, development of wetlands, planting and financial subsidies.
- Regulation setting rules for the amount of fertiliser and animal wastes that can go onto the land.
- Combination of measures and targeting of priority areas (presently considered the best option).

Public feedback and community input, particularly from the agricultural sector, to the regulatory group has been essential in this process. However it is essential that the benefits to the lake of any restrictions on agricultural activity are documented through appropriate monitoring to demonstrate goals and targets are achieved.

Solutions that are presently being considered include:

- Sheep and beef farming (decrease stocking rates, improve performance, mitigation)
- Dairy farming (decrease stocking rates, improve performance, mitigation)
- Urban and lifestyle development (upgrade septic tanks, stormwater control). New subdivisions will be encouraged to use new treatment technology.
- Industry and municipal discharges (upgrade sewage treatment, stormwater control)
- Forestry (little change)

Community and Government Involvement

A number of groups and agencies have been involved throughout the process of setting up management plans for Lake Taupo. A local interest group which has been particularly effective is the Lakes and Waterways Action Group (LWAG) which was involved early on in identifying community values and includes in its membership laypersons and scientists. Stakeholder groups such as local farmers and the tourist industry have become involved as options and solutions are sought. The two key groups in the development of the management plans are the local indigenous community and local and central government.

Indigenous community

"Taupo nui a Tia- The shoulder cloak of Tia". Since time memorial indigenous groups have been at one with the land and waters and in the case of Lake Taupo this can be traced back through local legends and mythology. The name of the local tribe, for example, can be traced through their great ancestor Tuwharetoa who is said to have descended directly from Ranginui the sky father.

In Maori mythology before there was any light there was only darkness, all was night. In this endless night Ranginui, the sky, dwelt with Papatuanuku, the earth, and was joined to her, and land was made. The children of Ranginui and Papatuanuku, who were very numerous lived in the darkness. At length the offspring of Ranginui and Papatuanuku, worn out with continual darkness separated the two. After failed attempts by some of the children it became the turn of Tanemahuta. Slowly, slowly as the kauri tree did Tanemahuta rise between the Earth and Sky they began to yield and then there was the world of light.

The numerous children that dwelled in the world of the light were the guardians and gods themselves the great ancestors of all parts of this new world. It is through this intricate web of genealogy that trees, birds, insects and finally humans were created.

Tia the great explorer who came to New Zealand from Hawaiiki on the Te Arawa canoe named Lake Taupo itself. Tia travelled inland from Maketu on the east coast where the canoe landed. He came across a large inland sea and erected a shrine on the eastern shores of the Lake at Hamaria and hung his cloak (taupo) on a post there. The lake and the area became known as Taupo nui a Tia, the great shoulder cloak of Tia.

Ngati Tuwharetoa see the environment holistically with all elements interwoven and intricately connected. Over the last millennium Ngati Tuwharetoa have developed tikanga (customs) and kawa (etiquette) that reflects this relationship with the environment. This knowledge and experience is fundamental to decision-making and the implementation of sustainable development for Taupo nui a Tia and its catchment (2020 Project team, 2002).

Lake Taupo and its tributaries were vested in Ngati Tuwharetoa through a Deed signed by the New Zealand Govt (Crown) in 1992. The trustees of the Lake for the tribal owners are the Tuwharetoa Maori Trust Board. The trustees are responsible on behalf of the owners for the day-to-day decisions required for the sustainable development of the Lake. Under the 1992 Deed the Minister of Conservation and the Tuwharetoa Maori Trust Board (TMTB) set up the Lake Taupo Management Board. The Management Board reviews resource consents pertaining to the bed of the Lake and advises the Trust Board of the implications of the activity. The other roles of the Management Board can be viewed through their Strategic Plan (Lake Taupo Management Board Strategic plan 2000-2005). As the landowner through the Resource Management Act (RMA) 1992 the final decision for development on the Lakebed still rests with the trustees for the landowners Tuwharetoa Maori Trust Board. Consent authorities must also consult with the TMTB before granting consents for users of water in the lake and its tributaries.

Key issues for the management of the Lake by the hapu (sub tribe) have been identified through the Tuwharetoa Strategic Plan (2000). This plan is owned by the hapu and was developed through the Environmental unit of the Trust Board through 2 years of consultation. The plan broadly states the key resources managed by the hapu and issues associated with them and some keys actions.

To build on the strategic plan, the Trust board have through the "2020 Taupo nui a Tia" project for Lake Taupo and Catchment developed an Environmental plan for the Lake. This detailed plan discusses policies and principles, which form a basis for protecting the taonga (resources, treasures) of the Lake.

Local and central government involvement

In New Zealand, natural resources are managed according to the provisions of the Resource Management Act 1991. The Resource Management Act, introduced in New Zealand in 1991, was the first legislation in that country to adopt an integrated and holistic approach to the sustainable management of resources, such as lakes, and replaced some 50 different pieces of legislation. Any activity which may have an impact on the natural resources of lakes must have an assessment of effects which is considered during consent hearings by regional councils.

Through this Act, central government has devolved the authority to manage land use in order to control the water quality of lakes and other waterbodies to two subsidiary levels of government. These are (1) Regional Government (as "Regional Councils"), ("... The control of the use of land for the purpose of ... the maintenance and enhancement of the quality of water in waterbodies ..."), and (2) Territorial Authorities (i.e. City and District Councils), ("... The control of any actual or potential effects of the use, development, or protection of land ...").

In the case of Lake Taupo, this means that both the Waikato Regional Council (also known as "Environment Waikato") and the Taupo District Council have the authority to manage land use in the catchment to maintain or enhance the lake's water quality. The councils have agreed that it is appropriate at present for the regional council to take a lead role, and for the district council to support this. As a result, the regional council has recently begun an intensive process of consulting with affected and interested parties. This is expected to result in a change—formally called a "variation"—to the regional council's existing "Waikato Regional Plan", a legally-binding document prepared under the RMA.

Future of Lake Taupo

Reversals of declining water quality in large lakes are achievable through active management of nutrient loads, as has been demonstrated for Lake Maggiore in northern Italy where oligotrophication followed a period of eutrophication (Ruggiu et al. 1998, Aldo et al. 2002). While it is hoped that the measures being put in place for Lake Taupo will form the basis for sustainable development there will need to be a major commitment in time and funding from all levels of society, industry, stakeholders, local and central government. It is only through such partnerships that targets such as those set for Lake Taupo and other lakes such as Lake Biwa will be met. The first step is acknowledgement of the problem and a commitment by all involved but the real success can only be judged through appropriate monitoring over the next 20-50 years.

Summary

Community values and specific local issues and circumstances are usually the main drivers of lake management. Because of these different drivers, the range of lake types and the social and economic structures particular to a lake, then different approaches are required. This process involves identification of issues and values, setting goals, options and strategies then monitoring to demonstrate they are effective. Involvement of indigenous groups, scientists, local citizen and communities, industry, end-users and stakeholders is essential for effective management.

Local lake management initiatives can be incorporated into national and global goals but are often ultimately driven by local issues. Voluntary changes and initiatives can be effective but legislation is usually essential and can be achieved at the local level in terms of regulatory frameworks. Broad objectives can be set at larger scales but global consensus and acceptance for water management is extremely difficult to achieve. Much can however, be learnt from other governments and communities who have faced similar problems through improved dialogue. The real success stories with lake management have been where good partnerships have been established between state and central government with all parties but the real action still needs to happen locally.

Acknowledgements

Thanks to Julie Hall and an anonymous reviewer for reviewing early drafts of this chapter.

References

ABIYA, I. 2001. Awareness creation and partnerships towards the management of lakes in Kenya. *In* Proceedings of the 9th International Conference on the Conservation and Management of Lakes, Otsu, Japan: 364-367.

ALDO, M., A. LAMI, S. MUSAZZI, J. MASSAFERRO, L. LANGONE, AND P. GUILIZZONI. 2002. Lake Maggiore (N.Italy) trophic history: fossil diatoms, plant pigments, chironomids and comparisons with long-term limnological data. Quaternary International, in press.

ANON, 1990. Levin scheme wins IPENZ Environmental Award. New Zealand Engineering **45:** 23-26.

_____. 2002. Tourism trends in the Taupo District. Destination Taupo, Taupo, New Zealand

BOPCC, 1975. Upper Kaituna catchment control scheme. Bay of Plenty Catchment Commission, Whakatane, New Zealand.

DONAHUE, M. J. 1996. A new era for regional water resources management: a Great Lakes case study. The 1996 Wayne S Nichols Memorial Fund Program, The Ohio State University, Columbus, Ohio.

ELLIOTT, A. H., AND M. J. STROUD. 2001. Prediction of nutrient loads entering Lake Taupo under various landuse scenarios. NIWA Client Report EVW01224. National Institute of Water and Atmospheric Research, Hamilton.

GOMEZ, Y. B. 2001. Watershed management advocacy and awareness building at the local level: the experience of the Philippine Watershed Management Coalition (PWMC).

In Proceedings of the 9th International Conference on the Conservation and Management of Lakes, Otsu, Japan. Session **2:** 377-380.

HADFIELD, J., D. NICOLE, M. ROSEN, C. WILSON, AND U. MORGENSTERN. 2001. Hydrolgeology of Lake Taupo catchment—Phase 1. Environment Waikato technical report 2001/01. Environment Waikato, Hamilton.

HALL, J. A., G. W. PAYNE, AND E. WHITE. 2002. Nutrient bioassays on phytoplankton from Lake Taupo. NIWA Client Report EVW01229. NIWA, Hamilton.

HEHANUSSA, P. E. 2001. Changing perspective of Indonesian water law, its relevance to lake management and conservation. *In* Proceedings of the 9th International Conference on the Conservation and Management of Lakes, Otsu, Japan. Session **5:** 357-360.

HOWLAND, W. G. 2001. Lake Champlain Basin Program: the structure of a model watershed partnership. *In* Proceedings of the 9th International Conference on the Conservation and Management of Lakes, Otsu, Japan. Session **5:** 618-621.

KORTH, R. M. AND L. L. KLESSIG. 1990. Overcoming the tragedy of the commons: alternative lake management institutions at the community level. Lake and Reservoir Management **6**(2): 219-225.

KURTYCZ, A. 2001. Citizen participation and the communication dynamics. The case of Mexico. *In* Proceedings of the 9th International Conference on the Conservation and Management of Lakes, Otsu, Japan. Session **2:** 373-376.

LWAG, 1999. Lake Taupo Accord (draft). Lakes and Waterways Action Group, Taupo, New Zealand.

MORGENSTERN, U. 2001. Age interpretation of tritium data from Mapara Stream water. IGNS Report to Environment Waikato. Institute of Geological and Nuclear Sciences, Wellington.

MWIVA, R. N. 2001. Environmental restoration of Lake Victoria. *In* Proceedings of the 9th International Conference on the Conservation and Management of Lakes, Otsu, Japan. Session **5:** 162-165.

OLIVER, J. 1995. Is the "limits of acceptable change" concept useful for environmental managers? A case study from the Great Barrier Reef Marine Park. *In* G. C. Grigg, P. T. Hale and D. C. Lunney [eds.], Conservation through sustainable use of wildlife. Centre for Conservation Biology, university of Queensland, Brisbane, Australia.

Organising Committee of the 9th International Conference on the Conservation and Management of Lakes Society and the water environment of Lake Biwa and the Yodo River Basin. 2001. Brochure edited and published by the Organising Committee of the 9th International Conference on the Conservation and Management of Lakes, Otsu, Japan.

RUGGIU, D., MORABITO, G., PANZANI, P., AND A. PUGNETTI. 1998. Trends and relations among basic phytoplankton characteristics in the course of the long-term oligotrophication of Lake Maggiore (Italy). Hydrobiologia **369/370:** 243-257.

RUTHERFORD, J. C., PRIDMORE, R.D., AND E. WHITE. 1989. Management of phosphorus and nitrogen inputs to Lake Rotorua, New Zealand. Journal of Water Resources Planning and Management, ASCE, **115:** 431–439.

_________, R. B. WILLIAMSON, AND A. B. COOPER. 1987. Nitrogen, phosphorus, and oxygen dynamics in rivers. *In* A. B. Viner [ed.], Inland waters of New Zealand. DSIR Bulletin 241. Department of Scientific and Industrial Research, Wellington. p. 139–165.

SCHOUTEN, C. J., TERZAGHI, W., AND Y. GORDON. 1981. Summaries of water quality and mass transport data for the Lake Taupo catchment, New Zealand. Water & Soil Miscellaneous Publication **24.** Ministry of Works and Development, Wellington.

SPIGEL, R., C. HOWARD-WILLIAMS, M. JAMES, M. M. GIBBS. 2001. A coupled hydrodynamic-ecosystem study of Lake Taupo: a preliminary model. NIWA Client Report CHC01/52. NIWA, Christchurch.

TUWHARETOA STRATEGIC PLAN. 2000. unpub held by Tuwharetoa Maori Trust Board, Taupo, New Zealand.

UTTORMARK, P. D., AND M. L. HUTCHINS. 1980. Input/output models as decision aids for lake restoration. Water Resources Bulletin **16:** 494–500.

VANT, B. 2001. Changes at Lake Taupo: the early warning signs? New Zealand Limnological Society 2001 Conference Abstracts, p. 21.

VANT, B. AND B.HUSER. 2000. Effects of intensifying catchment land-use on the water quality of Lake Taupo. Proceedings of the New Zealand Society of Animal Production **60:** 261-264.

VANT, W. N. 2001. New challenges for the management of plant nutrients and pathogens in the Waikato River, New Zealand. Water Science and Technology **43:** 137–144.

VANT, W. N., AND B. W. GILLILAND. 1991. Changes in water quality in Lake Horowhenua following sewage diversion. New Zealand Journal of Marine and Freshwater Research **25:** 57–61.

WATZIN, M. C. 2001. Developing ecosystem indicators and an environmental score card for the Lake Champlain Basin Programme. *In* Proceedings of the 9th International Conference on the Conservation and Management of Lakes, Otsu, Japan. Session **4:** 433.

WHITE, E., AND G. W. PAYNE. 1977. Chlorophyll production, in response to nutrient additions, by the algae in Lake Taupo water. New Zealand Journal of Marine and Freshwater Research **11:** 501–507.

________, ________, S. PICKMERE, AND P. WOODS. 1986. Nutrient demand and availability related to growth among natural assemblages of phytoplankton. New Zealand Journal of Marine and Freshwater Research **20:** 199–208.

Chapter 7

Global and Local Approaches to Freshwater Management: The Way Ahead

Warwick F. Vincent[1] and Michio Kumagai[2]

[1]*Dépt de Biologie, Université Laval Québec, QC G1K 7P4, Canada*
[2]*Lake Biwa Research Institute 1-10 Uchidehama, Otsu, Japan*

Abstract

There are compelling reasons for adopting a local perspective on water quality issues. Each lake or river has its own particular set of biological, chemical and physical properties and will therefore respond in a unique way to control measures. General models such as empirical relationships can be highly misleading if applied to individual sites without attention to these local circumstances. Environmental management strategies must also be integrated with the cultural, political, economic and regulatory aspects that are specific to each site, and local community involvement is an essential part of these strategies. However, an approach restricted to only local considerations is inadequate in the long term. High quality fresh water is fast becoming a high-value global commodity and will require global strategies for its preservation and for the regulation of its international trade. General (global) models, standardised methodologies (including paleolimnological analysis) and water quality data bases and criteria also provide a powerful starting point for local management strategies. This approach allows local managers to benefit from the combined pool of international expertise and research findings, to reduce the need for costly duplication of effort, and to sharpen research and monitoring objectives that will aid local decision-making.

Future management strategies will benefit from this global approach that is fine-tuned to local circumstances.

Introduction

The 21st century is likely to be a period of accelerating global demand for high quality fresh water, but will also be a time of increasing impacts of agricultural, industrial and urban activities on the aquatic environment. Lakes, rivers and wetlands account for only 0.3% of the total global freshwater reserves, yet they are the habitats for aquatic wildlife and are the main sources of water for human needs (Kalff 2002). The ongoing management of these natural aquatic ecosystems will require a diversity of approaches (Horne and Goldman 1994; Dodds 2002; Wetzel 2001) to meet the rapidly growing and conflicting pressures on this vital resource.

In this volume, specialists from a range of research backgrounds in the aquatic sciences have considered two distinct types of strategy to address future needs in water supply, monitoring and management. The first paper within each of the chapter themes has argued the value of general strategies that can be applied at sites throughout the world, and the importance of global initiatives. These include the development of a global data base for freshwater resources; the need to address global-scale impacts such as water shortages, exotic species invasions, climate change and long range contaminants; the development of standardized protocols for water quality monitoring and management; and the utility of generic approaches towards lake rehabilitation such as loading control and biomanipulation. The companion papers within each theme have placed greater emphasis on the need to consider local circumstances in terms of biogeochemistry, habitat variability, climate, economy and culture, and local scale protocols such as the use of wetland plants and algae in decontamination. This concluding chapter provides a synthesis of these perspectives and examines the strengths, weaknesses and complementarities of the two approaches that will help refine future strategies for sustainable use of the world's fresh waters.

Global Strategies

An improved understanding of the structure and functioning of lake ecosystems and their surrounding catchments is the foundation stone for long term stewardship of the world's freshwater resources. One of the impediments to such understanding at present is the limited opportunities for transfer of expertise between the separate subdisciplines of freshwater science (limnology) such as hydrodynamics, hydrology, bio-optics, microbial ecology, hydro- and biogeochemistry, production ecology, food web dynamics and landscape ecology. In this regard, the oceanographic research model is attractive in which many disciplines participate in a common sampling campaign from the same ship. Kumagai and Vincent (Chapter 1) provide an example of such an approach applied to Lake Biwa, Japan, within the program 'Biwako Transport Experiment'. In a subsequent program 'Cyanobacteria Risk Assessment at Lake Biwa', investigators from several speciality backgrounds took an ecological risk assessment approach to cyanobacteria blooms, and this led to important new insights into the biotic-physical relationships between bloom development, lake currents and mixing.

An essential step in monitoring the state of the world's freshwater resources is the creation of an appropriate information base that is accessible to all stakeholders. As noted by Robarts (Chapter 2.1), a United Nations assessment of the state of the hydrosphere at the end of the 20th century was hampered by a lack of reliable data from many countries and by the difficulties in comparing information from different countries and organizations using disparate types of data collection and reporting formats. To address this deficiency, the UN created the program GEMS/Water in which nations are encouraged to contribute to a common data base. This also links to The Global Runoff Data Centre in Koblenz, Germany, which operates a database of freshwater flows under the auspices of the World Meteorological Organization. Such information will be especially critical in determining how climate change is influencing the supply of water in different regions of the world.

Water quality and supply information is vital for meeting the resource needs of domestic, industrial, navigation, energy and agricultural activities, and it is also a key requirement for monitoring the habitat requirements for wildlife. The quality and quantity of fresh water exert an overall control on aquatic ecosystem properties, not only in lakes and reservoirs but also on coastal marine environments that depend upon freshwater inputs. An example is the huge annual input of sediment to the sea from the Yellow River, China, which is highly correlated with total discharge (Chapter 2.1). Similarly in many estuaries, the freshwater-saltwater transition zone is a region of high turbidity but also elevated biological production at all trophic levels, and these properties are strongly dependent upon discharge (Vincent and Dodson 1999).

Global data bases suffer from all the limitations of data collection across regional and national boundaries. Different laboratories often employ markedly different protocols for the same water quality measurements, and there can also be major differences in the quality of data. GEMS/Water is attempting to encourage standardization and cross comparisons, and is also involved in training initiatives to address these issues. Data are still sparse from many regions (e.g. the African continent) and this remains an impediment to an overall assessment of the state of the world's water. Even in developed nations, the monitoring sites and variables are not always representative of overall regional trends, or even of specific ecosystems.

Many questions in water quality management can only be adequately addressed by examining recent trends in ecosystem properties relative to the historical conditions that existed prior to human settlement. For most lakes and rivers, however, such long term data are sparse or (more typically) completely lacking. One of the few generic approaches towards filling this gap is that based on paleolimnological analysis of lake or river sediments. By way of examples from Québec, Canada, Pienitz and Vincent (Chapter 2.3) illustrate how an analysis of fossil diatoms in the sediments can be used to identify periods and causes of environmental change in the past. In Lake St-Augustin, this analysis showed how the current shift towards eutrophic conditions began more than 200 years ago and then rapidly accelerated with intensification of agriculture followed by the substantial increase in residential populations. In the drinking water reservoir Lake St-Charles, a change in trophic status coincided with the raising of water level and wetland flooding in the 1930s to increase the water supply for Québec City. The paleo-record provided no support for current concerns about deteriorating water quality in recent years and in fact indicated relative stability in trophic status. Paleolimnological analyses of trace metals in the sediments, however, indicate an ongoing rise in the concentration of these toxic substances, and argue against complacency in the ongoing management of this important urban freshwater resource. Paleolimnological analysis of the sediments of fluvial lakes in the St Lawrence River ecosystem has provided compelling evidence of recovery in water quality as a consequence of various pollution control strategies, however these analyses also show that much more effort is required to return towards pre-industrial conditions.

There is an increasing need for the formulation of global strategies towards the protection and restoration of lake ecosystems. In Chapter 3.1, Murphy describes a series of water quality issues that have resulted from the development of our industrial society and that now have impacts throughout the world. The increasing need for water in industry and farming has led to major engineering schemes, water diversion and aquatic habitat loss at many locations. One of the most extreme examples is the Aral Sea where large-scale irrigation to service the needs of the cotton farming industry caused a massive reduction in the inflowing river and a 64 % reduction surface area of the lake (Horne and Goldman 1994; Kalff 2002). Similarly, the diversion of inflows to Mono

Lake, California, to meet the burgeoning water needs of Los Angeles led to a shift towards a negative water budget for the lake. This in turn caused a massive reduction in water level and an increase in salinity that threatened the aquatic food web supporting several hundred thousand migratory birds which use this lake as a breeding site. As a result of a landmark legal decision, this large diversion has now ceased and water levels have begun to stabilize at previous values. As Murphy points out, the depletion of local water resources at some locations has sometimes resulted in a switch to alternative technologies, for example groundwater pumping and desalinization of seawater, each with its own new set of attendant problems in environmental management.

Long range transport of biota and contaminants continues to pose a set of problems for lake restoration and management that transcend national boundaries and local control strategies. The invasion of exotic species and the redistribution of the world's aquatic flora and fauna, for example via ship ballast, the aquarium hobbyist trade and aquaculture, has been a special problem for lake management in many parts of the world. The littoral zone of lakes such as Lake Biwa, Japan, and Lake Taupo, New Zealand, has drastically changed as a result of the arrival of aquatic macrophytes from North and South America, Europe and Africa. The invasion of zebra mussels from Europe swept through North America in the 1990s and resulted in the displacement of native bivalve species as well as the restructuring of some lake and river ecosystems. In the downstream reaches of the Saint Lawrence River, for example, the summer zooplankton community shifted from a community dominated by calanoid and harpactacoid copepods to one now dominated by the larval (veliger) stage of zebra mussels. The long term strategies for control of such invasions can only be met by global standards and international accords.

Persistent organic pollutants, pollution by heavy metals such as mercury, acid rain and several other pollution mechanisms require control actions at their industrial sources that lie up to thousands of km away from the sites of ecological impact. Murphy (Chapter 3.1) illustrates this effect by the example of sulphur loading in which long range transport of this element can reduce the ability of lake sediments to retain phosphorus, thereby accelerating eutrophication. The reduction of this and similar impacts can only be achieved through international cooperation in research to identify the mechanisms of effect and local remediation options, and through international consensus and development of regulatory policies.

The perturbation of biogeochemical cycles through human activities also has wide-ranging impacts in the aquatic environment. Eutrophication caused by agricultural phosphorus enrichment can be seen in this light as a global problem caused by the redistribution of the world's phosphate reserves to enhance crop production, with severe effects on the ecosystem structure and dynamics of downstream receiving waters (e.g., Lavoie et al. 2002). Murphy (Chapter 3.1) also notes the enrichment effects of transboundary fluxes of ammonia in promoting toxic algae such as *Microcystis*. The control of these and other harmful micro-organisms in drinking water supplies requires the ongoing international exchange of expertise, protocols and other information about the ecological mechanisms that influence the dynamics of these organisms, the best methods to monitor their biomass and toxicity, and the efficacy and human health effects of different control strategies.

A powerful approach among the global-level responses to water quality issues is the development of generic models that can then be fine-tuned to local circumstances. Legendre (Chapter 4.1) describes the utility of such models in which biological variables such as phytoplankton or bacterial respiration can be predicted from more easily observed environmental variables such as temperature and nutrient stocks. Generic models can be constructed at various levels of complexity and have the advantage of broad application to homologous systems. Such models can provide a valuable starting point to evaluating water quality issues, although as illustrated elsewhere in this volume (Chapters 4.2, 5.2) they may produce results that are inaccurate or misleading if distinguishing features of the local environment are not taken into account.

Eutrophication continues to be a global problem for drinking water supplies and the control of phosphorus loading is the generic method of choice to reduce and ultimately reverse the effects of enrichment. In their review of this approach, Jeppesen et al. (Chapter 5.1) note that although lakewater quality often improves in response to loading reductions, there are many examples in the literature of lakes that have shown a delayed or even lack of response. Such disappointing results can be attributed to a variety of causes including insufficient P reductions, continued enrichment by internal loading, and the effects of fish, both on sediment disturbance that increases P release and on reducing the populations of zooplankton, thereby reducing the rates of removal of algae by grazing. The latter effect lends itself to biomanipulation strategies whereby the fish community is manipulated to reduce the abundance of planktivorous species and allow high stocks of zooplankton. There is evidence of long term beneficial effects of this approach at TP levels below 0.05-0.1 mg P l^{-1}, although it may be less successful in warm tropical and subtropical lakes.

High quality fresh water must be increasingly viewed as a high value, global commodity, although there is also much concern about the implications of corporate ownership and control of this resource (Barlow and Clarke 2001). By way of examples from Lake Tahoe, USA, and Lake Baikal, Russia, Goldman (Chapter 6.1) describes the importance of long term commitments to limnological research, monitoring and management. Short-term data from Lake Tahoe, for example were touted as evidence that lakewater conditions had improved and that ongoing remediation methods were no longer needed. Subsequent monitoring showed that these years of data were simply momentary deviations from the long term trend showing continuous and ongoing deterioration in lake water quality, especially water color and transparency. Goldman draws attention to the power of the Secchi disc as a simple yet remarkably effective water management tool around the world to not only monitor a key aspect of the lake environment, but also to communicate water quality trends to the public and to politicians. This political engagement is essential for the long term protection of the world's fresh waters.

Local Strategies

The challenge of obtaining representative data for the long term monitoring of lakes is addressed by Frenette and Vincent (Chapter 2.2) who argue that while limnological tradition favours offshore pelagic measurements, there are often greater water quality concerns in the inshore littoral zone. The latter tends to be highly heterogeneous in space and time, thereby further discouraging routine measurements. Yet this heterogeneity is critically important to define and monitor for local water management decisions. For example, at Lake Biwa the offshore development of toxic blooms of cyanobacteria in the North Basin is dependent in part of the development of cyanobacteria at inshore sites under locally enriched conditions (Ishikawa et al. 2002). These nutrient replete cells are then advected offshore where they can continue to grow on stored reserves. The mapping of inshore sites of bloom propagation would help identify priority sites among bays, shorelines and sub-catchments of the lake for efforts to improve water quality.

One approach towards addressing the under-sampling of inshore sites is the application of bio-optical technologies such as surface colour observation systems (hyperspectral reflectance), *in situ* turbidity or fluorescence moorings, and spectral profiling of underwater light. The latter approach applied to Lake Biwa, Japan, has underscored the large site-to-site variations in the underwater exposure to damaging UV radiation as well as the variability in energy supply for photosynthesis (photosynthetically available radiation, PAR) and in waveband ratios that may cause differences in UV damage and phototrophic growth rates between species. A particularly striking example of local variability in limnological conditions is fluvial Lake St-Pierre in Québec, Canada, where source waters from different catchments differ greatly in their concentrations of CDOM (coloured dissolved organic matter) that in turn results in visually distinct, bio-optical differences between sites. This degree of variability underscores the need to monitor at a suite of sites and precludes making any general inferences about changes in the lake environment based on measurements from a single location. Each lake is different and requires a monitoring strategy that addresses its particular scales of variability.

As a counterpoint to the global perspective on controlling enrichment and pollution effects in the aquatic environment (Chapter 3.1), Peterson considers the opportunities for site-specific remediation using technologies based on local wetland plants and algae (Chapter 3.2). This approach can offer advantages over chemical stripping techniques that result in undesirable end products, and engineered microbial systems that require meticulous control of energy and nutrient inputs. Managed wetland systems for waste water treatment offer the advantage of low cost in combination with high capacity for nutrient stripping and decontamination, and habitat space for aquatic wildlife including birds and amphibians. However the efficacy of such systems varies greatly, and is a function of many local variables including climate, hydrology and plant species composition.

In many parts of the world, drinking water is stored in shallow reservoirs that are subject to a variety of water quality problems including nuisance algal bloom development. Such waters are often managed on a local, *ad hoc* basis in which quality control is only performed in response to specific problems and crises. Peterson (Chapter 3.2) provides a compelling example of the inadequacies of this approach. A drinking water reservoir was treated with multiple doses of copper sulphate to suppress a bloom of potentially toxic cyanobacteria of the genus *Aphanizomenon*. This treatment, however, resulted in toxic concentrations of copper, destruction of the zooplankton community and a shift towards high populations of another group of nuisance algae, minute *Chlorella*-like cells that proliferated in the absence of zooplankton grazing and that were too small to be removed by the water filtration plant. Peterson advocates the application of milder treatments to minimize such deleterious side effects in drinking water supplies, and provides a classification of control agents that act with different levels of ecological impact. This example argues the need to consider global criteria in the selection of control measures as well as the importance of adopting longer term strategies to ensure the supply of high quality drinking water. These 'whole ecosystem' strategies ultimately depend not only upon in-lake monitoring and treatment but also, and most importantly, on the successful management of diffuse and point sources of enrichment, contaminants and pathogenic microbes from the surrounding catchments.

There is a vital need for local data and an understanding of the particularities of individual ecosystems in applying numerical or other types of models to local water quality issues. Yamamuro (Chapter 4.2) illustrates this point with her description of a biogeochemical model for Lake Nakaumi, Japan. This shallow water system is poorly described if only pelagic processes are considered because of the major role played by high populations of a filter-feeding bivalve. This species exerts a strong effect on the standing stocks of phytoplankton in the overlying water column, and is therefore an essential component of water quality models for this lake. The experiences from this lake and adjacent Lake Shinji also underscore the importance of adequate field data from the local environment to set the model parameters within realistic bounds.

The theme of local understanding is further developed by Howard-Williams and Kelly (Chapter 5.2) who argue that the choice of appropriate lake management and restoration strategies should be mostly dictated by the specific properties of the local site. Three sets of environmental properties require particular attention. Firstly, catchment geology influences the magnitude and nature of nutrient limitation via the geochemical weathering of rock in the surrounding catchment. The importance of these natural sources of nutrients will influence the success of nutrient control measures, and the need to control N as well as P loading. Geomorphology and other landscape properties also affect groundwater characteristics and can result in legacies of earlier periods of enrichment that extend beyond the hydraulic residence time of the lake. Secondly, local climate influences the stratification and mixing regime which in turn regulates the timing of events that affect biological processes and water quality. Finally there are biogeographical considerations such as the presence of herbivorous fish communities and the arrival of invasive plant and animal species.

The limnological and lake management literature tends to be highly biased towards European and North American aquatic ecosystems, and this information may be poorly applicable to lakes elsewhere. An example is the N-limited lakes of the central North Island, New Zealand, where the catchments are naturally rich in phosphorus and management strategies based exclusively on P control than on N+P loading are likely to have little success. The combination of regional climate and nutrient regime for these lakes results in unusual seasonal patterns, with maximum algal biomass and primary production occurring during winter. Similarly the warm, ice-free conditions of these lakes allows macrophyte growth throughout the year. These distinct local characteristics mean that some types of generic models such as regression equations based on studies elsewhere are likely to be a poor guide, and even highly misleading, for decisions concerning local water quality management. Similarly, management protocols based on the north temperate experience (e.g., a single annual cut of macrophytes to control their growth; biomanipulation without considering the specific characteristics of native fish and zooplankton communities) can be ineffective and inappropriate as local control strategies unless they are modified to account for site-specific properties of the lake and river ecosystem. This similarly argues for the critical importance of an advanced limnological understanding of local ecosystems.

James et al. (Chapter 6.2) describe the vital importance of community involvement in successful lake management. These authors note the challenge faced by lake managers who must balance the needs of current lake users with long term protection of the resource, all while ensuring an open dialogue with all stakeholders and minimum conflict. The most successful long term planning and management is at a 'whole-ecosystem' catchment level, although such an approach has been lacking for many aquatic environments, for example some large river ecosystems (Vincent and Dodson 1999). James et al. identify five key features that will determine the success of an integrated management strategy: a detailed description of the resource; consensus on key issues and environmental values; the setting of appropriate goals and compromises; the exploration of management options; and finally a cost-benefit analysis of these options prior to implementation. The foundation of the approach is the resource description that incorporates the most recent scientific findings including information about the causal relationships between water quality and other characteristics of the lake and its watershed. Consultation with the full range of stakeholders is desirable throughout this process, including those representing present-day users and future generations who will benefit from careful stewardship of the resource.

James et al. cite the example of Lake Taupo, New Zealand, where a variety of conflicting demands for water are being addressed in the development of a long term management plan integrating social, cultural, environmental and economic knowledge. This clear-water lake is used for tourism, a variety of water sports including fishing and sailing, hydroelectricity and drinking water. The ownership of the lake and its tributaries lies with a tribe of indigenous people, the Ngati Tuwharetoa Maoris, and the waters and associated lands have great cultural and spiritual significance that must be addressed in all management decisions. Pastoral farming is an important activity in the surrounding catchment and current proposals to intensify these activities have generated concern about impacts on water colour, transparency, oxygen and other key properties of the lake. Via modelling analyses and extensive consultation with all local stakeholders, a series of options were identified and evaluated, and consensus has been achieved on striving for the long term preservation of the current high-quality waters of Lake Taupo via a variety of regulatory and non-regulatory means. Similarly at Lake Biwa, Japan, a 20–year plan has recently been initiated to restore water quality and to ensure the long term protection of the lake. This program similarly involves partnerships between local communities, government agencies and industry.

Conclusions

The arguments for a local perspective on water quality issues are compelling. The quality of water is of greatest interest to the members of the local community who use it and who will experience directly the positive and negative (including health and financial) consequences of environmental management decisions. As illustrated by the examples above, rehabilitation and protection of freshwater resources must take careful account of many local factors, ranging from regional climate and catchment hydrogeochemistry to cultural and spiritual values. However, an exclusively site-specific orientation has its limitations, and for some types of water quality issues, for example those involving long range transport of pollutants, such an approach is manifestly inadequate. Water quality monitoring and management at all sites can greatly benefit from the global sharing of information, sampling and analytical protocols including new advances in technology and understanding. Similarly there is enormous value in the sharing of experiences, both positive and negative, of applying different monitoring, protection and rehabilitation strategies, different responses to water quality problems, and different management structures and activities to ensure involvement by all stakeholders. International organisations such as GEMS/Water, the international society for freshwater ecologists (Societas Internationalis Limnologiae, SIL) and the International Lake Environment Committee (ILEC) have key roles to play in facilitating this exchange of information, expertise and experience. Of course the evaluation of such information must always be tempered with local scientific understanding and a critical analysis of its goodness-of-fit to local circumstances. Indeed, this comparative analysis in itself is likely to foster improved understanding and dialogue, and may help sharpen management objectives as well as gaps in understanding requiring targeted research.

The ongoing deterioration of aquatic ecosystems throughout the world at a time of accelerating demand for high quality freshwater heightens the urgency for improvements in lake and river management. As noted by Smol (2002), human society has a long history of underestimating its impacts upon the environment. In the aquatic domain we have seen many examples where rapid degradation has led to ecosystems that can only slowly be restored, and perhaps never to their original conditions. The forecasted major changes in world climate over this century will even further compound the challenges faced by water resource managers. Future management strategies will benefit from an openness to global protocols, accords and information sharing, while addressing the need to fine-tune to local circumstances.

References

BARLOW, M., AND T. CLARKE. 2001. Blue Gold - The Battle against Corporate Theft of the World's Water. The New Press, N.Y., 278p.

DODDS, W. K. 2002. Freshwater Ecology – Concepts and Environmental Applications. Academic Press, Ca. 569p.

HORNE, A. J. AND C. R. GOLDMAN. 1994. Limnology, 2nd ed. McGraw-Hill Inc, N.Y., 576p.

ISHIKAWA, K., M. KUMAGAI, W. F. VINCENT, S. TSUJIMURA, AND H. NAKAHARA. 2002. Transport and accumulation of bloom-forming cyanobacteria in a large, mid-latitude lake: the gyre-*Microcystis* hypothesis. Limnology **3:** 87-96.

KALFF, J. 2002. Limnology - Inland Water Ecosystems. 2002. Prentice Hall. 592p.

LAVOIE, I., W. F. VINCENT, R. PIENITZ, AND J. PAINCHAUD. 2002. Effect of discharge on the temporal dynamics of periphyton in an agriculturally influenced river. Revue des Sciences de l'Eau (in press)

SMOL, J. P. 2002. Pollution of Lakes and Rivers A Paleoenvironmental Perspective. Oxford University Press, USA. 280pp.

VINCENT, W. F., AND J. J. DODSON. 1999. The need for an ecosystem-level understanding of large rivers: the Saint Lawrence River, Canada-USA. Japanese Journal of Limnology **60:** 29-50.

WETZEL, R. G. 2001. Limnology : Lake and River Ecosystems, 3rd ed. Academic Press. 1006p.

Subject Index

GPSR Compliance

The European Union's (EU) General Product Safety Regulation (GPSR) is a set of rules that requires consumer products to be safe and our obligations to ensure this.

If you have any concerns about our products, you can contact us on ProductSafety@springernature.com

In case Publisher is established outside the EU, the EU authorized representative is:

Springer Nature Customer Service Center GmbH
Europaplatz 3
69115 Heidelberg, Germany

Batch number: 10584436

Printed by Printforce, the Netherlands